BLACK WATER

BLACK WATER

A Life in the Special Boat Service

Don Camsell

LEWIS

INTERNATIONAL, INC.

Published in the United States in 2000 by
Lewis International
2201 NW 102nd Place, 1
Tel: 305-436-7984/800-259-5962
Fax: 305-436-7985/800-664-5095

First published in Great Britain by
Virgin Publishing in 2000

A catalogue record for this book is available from
the British Library.

ISBN 1-930983-00-X

Typeset by TW Typesetting, Plymouth, Devon

Printed and bound in Great Britain by
Creative Print & Design (Wales), Ebbw Vale

Contents

To the men of the Special Boat Service: soldiers, corporals, sergeants, sergeant majors and officers of this unique Special Forces Group whose honourable profession I had the privilege to serve in for 22 years. This book is dedicated to you and especially those loved ones not with us today: Chris T, Chris D, Coke, Kiwi, Lou, Neil B and Dickey H, and to our extremely good friend Chris Brogan and his chair and the irrepressible Kevin J. I offer you all my respect and affection. I thank you all.

Acknowledgements

With thanks to General Sir Peter de la Billière, HarperCollins Publishers and Curtis Brown Ltd for kind permission to quote from *Storm Command*.

A special thanks to the Ministry of Defence for their kind co-operation.

Thanks to Will Pearson who laid the foundation and idea, to Dino Boni, and to Ian Drury who deciphered my writing and laid the project to rest.

A special thanks to my publisher, Humphrey Price, for his wisdom, his understanding, his dedication and expertise.

The biggest thanks of all to my wife Fiona and my family who supported me over the years within an organisation where the family, in most cases, came second. Your continued support is appreciated. I love you all.

Author's Note

Security considerations alone restrict what I can reveal about the Special Boat Service. I hope the book will lead the reader to ask himself how these men overcome the agonies and push themselves to the extremes of human endurance. Why do these men, trained to the highest standard of self-discipline and proficiency, offer their lives for their country so willingly. Most of the time, their families are not even a concern: it is the determination to become someone unique, to accept the challenge of a lifetime. Once fulfilled, the stimulus of this type of work is such that they want more. The more challenging the task, the greater the *esprit de corps*.

The names of some characters and places in this book have been changed, but in all other respects, the book is entirely factual. If parts seem extreme, it's because truth is indeed stranger than fiction. The achievements of this uniquely professional service are second to none.

Introduction

I WAS DYING TO PUT THE HEATER ON, but I couldn't risk them hearing me start the engine. Half the night we'd sat in my battered Datsun, watching her flat. She wasn't in, but she'd be along soon. In any case, we needed to see if anyone pitched up while she was away.

You can hunker right down in these old Datsuns. Unless someone walks right up to the windshield, they can't see if anyone's inside. If the IRA knew we were here at Warrenpoint, watching their quartermaster's every move, they'd probably take us out. They had a few M60 general purpose machine guns and a number of RPG-7 anti-tank rockets: enough to blow us away in an instant. Alternatively, if our parking spot was compromised, they could always pack the vehicle next to us with Semtex.

Such jolly thoughts kept us going till the first hint of dawn. The radio hissed into my ear. She'd left her family's pub, possibly with company, and was heading our way. We really ought to move to a more distant location now that it was nearly daylight, but we'd have a great view from here. I could barely bring myself to look around as curtains were drawn in neighbouring houses. Who was looking at us from behind those black windows?

Rain drummed on the car roof. A good Irish downpour. Water flooded along the gutter, a white spout soaring above a blocked drain. A funny moment to be reminded of what I was really trained for. I was a swimmer-canoeist in the Special Boat Service, the SBS. Trained for amphibious raiding, maritime counter-terrorism, demolitions, communications ... you name it, we did it on water. It took me the hardest physical five and half months that I can remember, but I conquered the 'Black Waters' – the secret name of the Special Boat Service selection course. Like the other successful candidates, I achieved this with the help of my Royal Marine Commando training. This was, and still is, the foundation of our imagination, personality, fitness and professional soldiering – something that no other Special Forces organisation can fall back on.

So why was I parked up a side street in Northern Ireland, a Browning 9 mm pistol hidden under my tatty civilian coat? I was seconded to 14 Intelligence Company, 'the Det', a unit recruited from all across the British forces. It was another almighty challenge, working with the Det. And I loved it.

Another Datsun pulled into the street. It was her. I'd shadowed the Bear Cub for seven months and knew every detail of her life. I'd even copied her car. She cached weapons and explosives for the IRA, transporting them to pre-arranged locations on the eve of a terrorist operation.

She did have someone with her, but I didn't get a proper look until he stepped out and stood under the porch while she locked the car door.

It couldn't be him. Not here. We thumbed through our photo album to check. No doubt about it: the Big Man, commander of one of the IRA's brigades. A face drummed into us back in barracks; we had no idea he was on our turf.

1 Black Water

I STEP ON TO THE PITCHING DECK of the submarine. Rain spatters against my facemask, the sea invisible but for white spray breaking over the bow. I pause to give my eyes a chance to adjust to the darkness. The submarine is pointed into the swell. The sea state looked well over force 5, anything beyond force 7 would be suicide.

Hunched figures emerge on top of the fin: the captain and duty watch of HMS *Orpheus* maintain 360-degree surveillance as the rest of the team follow me along the glistening casing. We make our way to the 'lurking area' by the forward torpedo hatch. We had been cooped up below for the past 36 hours. Time to earn our money.

There is just room for a team of divers to sit in the lurking area of the old 'O' class submarine, breathing from 150 cu. ft air bottles as the submarine makes it's final run in to the target. It's a slow business getting everyone in. Each man checks his fins, mask, hood and gloves. He also ensures that he has a 0–200 ft depth gauge and two breathing units from the 150 cu. ft bottles. The last check is to make sure that air is registered on the 150 cu. ft gauge.

The submarine crew who had helped us on deck scuttle back inside the fin. The hatch swings shut. The heads disappear from the top of the fin as the watch rattle down the ladder into the light and warmth of the control room.

Out here on the casing, the magic world of information technology has yet to come to our aid. The signal procedure between the submarine and the divers is very primitive. The crew use a series of knocks to indicate to the team what to do and we in turn reply the same way – 'taps and bells'. Last chance to check your gear. Cold, clumsy fingers clutch at the breathing units.

Three sharp taps from the hatch below. The sub is about to dive. I take a deep breath and shout, 'Everyone ready?' So far, so good. I motion to Rupert (the boss) that I'm about to give the OK. He gives me the thumbs-up. I bang on the hatch three times, the sound echoing around the confined area. Everyone tenses as the air hisses from the flotation tanks and the bow dips into the sea.

Black water laps around our ankles. I kneel in front of the hatch, eyes locked on to my depth gauge, my hammer in my hand, ready to pummel on the lid if anything goes wrong. Now the water is at our waists and waves pour over the top into the lurking area. The lurking area is engulfed by the sea and we breathe from our air bottles. The submarine continues its dive. My job is to ensure that the team is OK, and I study each individual in turn; but I never lose sight of the depth gauge. If the casing were to go below 50 feet and we were on a training mission I would give the signal to surface. In this case I doubt whether I would surface the boat unless we ended up much further down.

The sub levels off, the steady rumble of its screws clearly audible. We settle back behind the protective shield as the boat works up to 15 knots. The darkness is total now, and all movement is by touch alone. You sense rather than see the wake rushing past above the shield: raise your hand and it is whipped back down at once. For this reason we have to maintain negative buoyancy. Float up and you would be swept away in an instant. It seems an age, but with my eyes finally adjusted to full night vision I can just make out the silhouettes of the nearest divers. I clench and unclench my fists and roll my shoulders to keep my circulation going. Whatever you do, it's a cold, numbing ride. And we have a long hard climb ahead of us.

A glance at the watch tells me the submarine will slow down soon. Time to switch to the attack oxygen breathing set, the LAR 5. Using taps and bells we get the signal and conduct the changeover. Each man partners another as using oxygen at depth is dangerous. The air breathing unit is removed from the mouth and replaced by the oxygen breathing unit. You clear your oxygen set of any impurities and then conduct a two-minute test on oxygen, with your air breathing unit ready to hand should anything go wrong.

The team conducts this procedure in a sequence working from the port side (left) around to starboard (right). When everybody is on oxygen the signal comes back to me, above the forward torpedo-loading hatch. Then Rupert and I conduct the same procedure. Once the whole team is on oxygen and ready for release, I signal to the submarine again through a series of taps and bells. Every item of equipment that is not being used is stowed and secured under the casing.

The submarine slows down.

Four sharp taps from inside the sub. I point at the first pair of swimmers. They seem to move in slow motion, rising to move along the casing. They pass above us, grasp the lines, pause to check their compasses, then kick out with their fins. The two swimmers vanish into the murk. This is repeated until all the team have left the lurking area apart from Rupert and myself. We check that all the swimmers have departed and all equipment is stowed. I signal to the submarine that we are the last swimmers to leave and they can proceed at normal speed out of the area. We pull ourselves clear of the lurking area and feel the force of the current. I check the compass, confirm with Rupert that he is ready and we release ourselves from the submarine. I slightly inflate my life jacket to gain some buoyancy and then swim towards the target. I press my face into the compass board to get some indication of the time we have been swimming. I estimate 22 minutes and decide to surface and see how near we are to the target. I indicate my intentions to Rupert and rise from the deep.

Towering above us, 250 metres dead ahead, is the black mass of an oil rig. Waves break against its gigantic steel legs. The equivalent of a 20-storey building looms out of the sea. Far above, lights glimmer through the spray. I duck straight down again, swimming along the buddy line to where Rupert is waiting. We swim on the same compass bearing until there is a sudden surge of water and we are drawn in by the sea, rushing in and out of the rig legs. Somewhere around here should be the 'clothes line' that the first teams have secured under the rig legs. We eventually find it, discovering several sets of diving kit already attached. I wriggle free of my diving set and secure it to the line. With one final deep breath, I take out my mouthpiece, switch it off, and kick out for the surface.

Great surges of water make it impossible to maintain position. The sea is foaming white all around the rig. The wind-driven spray almost drowns out the sound of my radio, but I can hear that the helicopters are somewhere out there, an RAF Nimrod jet co-ordinating the attack from above. Rupert remains on the control channel while I switch to the team channel. The coded chatter is deliberate and professional, in total contrast to the crazy scene before my eyes. Two lines dangle down from the glistening metal above, swinging wildly in the wind. A tiny speck inches its way up the line: one of the pole men, over a hundred feet up in the air. The ladders tumble down to settle just above the surface of the sea.

The swell engulfs the ladders, then recedes to leave them eight feet or more above our heads. You wait for the waves to surge up, taking you with them, then grab the ladder and cling on as tightly as you can. I can see Simon, Alan, Paul, Dave and Chris directing the team on to the ladders. One by one, black silhouettes cut through the waves to reach the ladders. I recognise Charlie straight away. A small man who has a great future within special forces: fit, strong, witty and intelligent. Charlie grabs the ladder on the upward surge of the water, hanging there with only one arm. He reaches up with his second arm and hauls himself up, placing his feet and fins into the gate of the ladder. Suddenly, an enormous wave swallows a third of the ladder – and Charlie. It seems to stay there for ever before finally receding, leaving Charlie still attached and climbing up as if nothing has happened. The second ladder is now in the water and a second team member is on board and heading for the top.

The two lead climbers are about one ladder length up (25 feet – we had four attached together to make a 100-foot climb). At this point you stop and hook on: remove your fins and attach them to your belt. You wouldn't last long in a gun battle top side waddling around like Donald Duck. Charlie and Les take off their fins then race up the ladder. These guys have formidable upper body strength; even in the teeth of the storm, the ladder swaying out to 45 degrees off vertical, they power up the ladder. They reach the top. The rest of the team cluster round the foot of the ladders, struggling to keep their position in the surf. One by one, they clamber on and upwards to the cable deck so dauntingly high above our heads.

My turn. Rupert is already on the other ladder. I let a big wave lift me part the way up and grab hold of the ladder. The wave races away as quickly as it has risen, the water pulling at my legs, trying to take me back down. It is a relief to take the weight off my arms and shoulders for a moment, but I still have 80 feet to climb and the wind takes the ladder and swings me out at a sharp angle. The trick is to get into a rhythm and not give yourself a chance to think how much it hurts. And never think how long this will take: to get to the top of a rig in conditions like tonight can take two hours – even when you're really good at it. Don't overreach. Power up with your legs to spare your aching arms as much as you can.

About halfway up, I feel a sharp tug on the ladder from below: another team member starting his climb. I lose my rhythm and become painfully aware of my shoulders, burning with the effort and threatening to seize up. My life-jacket inflation line gets caught in the ladder and inflates the jacket: it takes every ounce of strength to hook on again and deflate the damn thing. Once you've stopped climbing it hurts like hell to start again and I'm carrying enough kit to start a small war: Heckler and Koch assault rifle strapped to my right side; Sig-Sauer 9 mm pistol on my right thigh; waterproof 'flash bang' stun grenades and water- and pressure-proof radios. When I finally go up the next rung my arms feel like I've been climbing all week. But I have no choice. I am dangling on a spindly ladder suspended between the legs of an oil rig. Below is an angry sea. Above? I'll find out when I get there.

North Sea Oil and Gas rigs became a priority to the Ministry of Defence (MoD) in the late 1960s. They were considered vulnerable to terrorist attack and the task of protecting them – and, if necessary, storming one in terrorist hands – fell to the SBS. Like multi-storey tower blocks incongruously slapped down in the middle of the sea, the rigs look even more massive from water level. Storming a 30-floor tower block is complicated enough. Doing it at sea takes it to another level. And we can never forget that these national assets are highly explosive too. The planning of a hostage rescue or all-out assault has to take into account the risk of stray shots blowing terrorists, rig workers and us into oblivion.

This operation began with an eight-hour drive, from Poole in Dorset to the nuclear submarine base at Faslane in Scotland. We were used to the routine here. A stony-faced constable plodded to the vehicle with measured steps. MoD police are famous for their great sense of humour, so when he leaned in at the driver's window, we ask him for four Big Macs, fries and milkshakes to go. Not really. We sat there poker-faced while Rupert dealt with the paperwork. It's a dismal place, surrounded by hills and the weather unchanging. The cloud base was on the floor, rain drumming on the car roof. The policeman stood by the car, water running off his cap, down the sides of his face and into his shirt collar.

Faslane is almost like a second home, we've been there so often, and yet the security was still as tight. Rupert and I wondered why we had been suddenly recalled and sent to Faslane as part of M Squadron SBS. Our job is the protection of maritime installations, fixed or afloat, anything that might be a target of a terrorist group. We could imagine many things, but we were superbly confident in the ability of this team. But what was our mission today?

At last, the remaining team vehicles arrived and we were waved through to 'red area', the maximum security sector where the submarines were berthed. The eight team and support vehicles slid through the narrow wharf ways under the cranes. We passed a number of nuclear submarines including HMS *Churchill*. The *Churchill* brought back exciting memories of the Falklands War. In 1980 we'd made the same drive at desperate speed, ready to pile ourselves and our kit into the submarine. Within 20 days we were 8,000 miles away, off the island of South Georgia. To this day the crew of *Churchill* and above all her captain are remembered favourably by all who participated in that operation. The rush of travelling at 30 knots at 600 feet does not happen every day – let alone for twenty days!

The naval personnel knew the sight of the blue team vehicles and pitched in to give a hand, even though the weather was crap and it was 10 o'clock at night. We eventually pulled up by a diesel-electric submarine, HMS *Orpheus*, which we had been aboard many times. Rupert, as the troop officer, was immediately summoned inside. He asked me to accompany him, but I decided to stay with the team. It showed Rupert and the team that I had every confidence in him: he didn't need me to support him at the briefing. We had been together for eighteen months and the team was like a well-oiled machine by now: everyone confident of themselves and the rest of the guys.

Rupert went down the ladder and we hung about, ready to board and rig the submarine. I was the sergeant major of the team, so the team leaders descended on my vehicle, demanding to know what was happening. All I could say was that Rupert was being briefed. Until then, sit tight.

The whole team clustered around the vehicle. Wind and rain sliced right through their clothing, but the guys were young, full of confidence and wild speculation about what we would be ordered to do. Who cared about the weather? Having been in this situation many times I was more relaxed but conscious of a certain tension, adrenalin pulsing through the system.

Rupert emerged from the fin door twenty minutes later. He looked the part: cool, professional – a far cry from the nervous young officer who had been assigned to me many months ago. You could sense that the team leaders, consummate professionals themselves, trusted the man and the decisions that he would make. His initial orders were short and sharp. The target was a gas rig in the Brent Field, north of Scotland, which had been hijacked by a terrorist group. The submarine had to sail by 0400 hours.

The aircraft carrier HMS *Invincible* was to be the main forward mounting base (FMB) and the rest of the squadron was already flying out to her. That gave us five and a half hours to rig the submarine. Our stated H hour (attack time) was 2200 hours, the following night. Rupert gave responsibility for rigging the submarine to Simon and the team wasted no time in getting down to business. We had all the support from the submarine that we required. We'd used this sub for a training exercise six weeks before; routine and liaison with crew was established.

We each had a job to do. No task, however minor, was too much trouble: proper planning prevents piss poor performance. Our personal dive equipment was top priority. We passed it down through the forward torpedo loading hatch. It took a couple of hours to assemble all the kit, but eventually the lines were in position from the lurking area by the bow to the top of the fin. With all the air bottles and team equipment in position, we went for a quick rehearsal. Having done this exercise many times with this team and boat crew, we conducted separate rehearsals before coming together for a full co-ordinated dry run of the multi-swimmer release. If something was to go wrong, we'd rather it happened now where we could sort it out, than in the middle of the North Sea with a rig full of terrorists sitting above us.

0400 hours: tugs shepherded the submarine from its berth and down the loch. The sub's diesel engines kicked in and HMS *Orpheus* made for the open sea. We would make the first part of the voyage on the surface; although *Orpheus* could make up to 17 knots submerged, she used electric motors when underwater and high-speed dashes drained the batteries very quickly. The captain would want to keep his batteries fully charged. No one knew how long we'd have to lurk unseen near the target. Staying on the surface also let Rupert and the boat's skipper keep in satellite contact with London and in radio contact with the rest of the forces involved in the operation.

Rupert told us that HMS *Invincible* was nearing the target area and would keep station 30 miles from the rig: just below the radar horizon from the rig. It would take our submarine twelve hours to reach the rig. That seems a lot of time, but we had individual briefings to give and kit to sort out: once in the sea there would be no chance to recheck. We did allocate time to rest, of course. I'd like to say that, like the professionals we are, we finished our briefings, checked and tested the gear then slept like logs so we arrived all fresh-eyed and bushy-tailed. Except that we were all keyed up, even if we didn't admit it to each other. You keep thinking, I did check my life jacket, didn't I? and rechecking the pressure gauges on the air bottles.

Have you got the correct pressure for the time you're expected to be under? One thing you can't check is whether you stored the equipment topside correctly. You'll find out when you surface: if it wasn't secured properly, it will have been washed off. Above all: do I know my drills? If my Heckler and Koch gets a stoppage, will I deal with it quickly enough not to endanger myself or my team? (The trick is to work out within a second whether you can clear it or whether you should drop it and draw your Sig-Sauer pistol – take any longer to make your mind up and you might be dead.) And if it does come to shooting, will the ammunition work? Anyway, too much to think about, so sleep proves impossible.

Alarm bells! Dive stations: HMS *Orpheus* inclined gently, the waters rising around her fin as the surface watch slipped down the ladder. Watertight doors clanged shut all around us. The distant rumble of the diesels stopped and it felt as if we were slowing down. Her electric motors were barely audible, but we were actually working up to nearly maximum speed: the captain had timed his approach to perfection.

Crammed around a diagram of a rig, we planned how to get from the water to the lowest part of the rig, a good hundred feet or more above sea level. And what to do if the opposition was there and waiting for us. Above all, where were they holding the hostages? We had to cut down the time between the terrorists realising they were being attacked and our teams finding the hostages. Each team leader received his objective: the control room, the restaurant, the flight deck . . .

I took another look at the team leaders on whom this would largely depend. Simon was 6 feet of Scottish chauvinism, oozing so much confidence that you wouldn't mind if he fell on his face this time – except he was too nice a guy. When in doubt, just take the piss about his receding hairline. Alan was another Scot, more flamboyant and better in the water, something he never let Simon forget. Chris was an Aussie, on exchange from their SAS regiment and probably the most talented practically professional man we had ever worked with. Fortunately for us, when he finished his two-year posting and it was time for him to return to Oz, he decided to leave the Australian SF and join the Royal Marines, Special Boat Service. Tragically, he was to be paralysed in an accident a few years later in Belize. Dave and Paul were the typically hardworking team leaders. Dave was the unit comedian, an overenthusiastic SF soldier who made mistakes, but always survived – just; he never missed the chance of a quick one-liner, even when it dropped him in it.

Rupert had five four-man teams; he and I made up the HQ team that would control the final assault. Also in support of us would be the emergency response (ER) who would be mounting from the forward operating base (FOB), in this case HMS *Invincible*. These teams had two definite missions. If the swim assault teams (us) failed to make it to the target, they would be prepared and briefed to take the priority targets. Secondly, they would become our back-up teams once the assault was under way.

There was one limiting factor that had to be considered during this planning. From the time that we were released from the submarine, swam the distance to the target, climbed and established a foothold on the target, we had 90 minutes. Everything to go like clockwork. I could not help but analyse how all our training had gone over the past few months. Had we done enough for this type of operation? Yes. We were the best.

We had trained to attack all sorts of targets in every kind of situation: ships under way, ships at anchor, ships alongside in harbour as well as the oil and gas rigs. Out of all these options this particular one was by far the toughest, mentally and physically. I had covered this task from the moment the mission was handed to the SBS. I had realised from the beginning that this was the most dangerous, demanding and rewarding job in the services.

Rupert scribbled away, putting together his appreciation that had to forwarded to London for approval before we could go firm on our proposed plan. He looked up at us from time to time, then it was back to the notepad: he was on his own here. Rupert seemed older than he really was, receding dark hair – perhaps it's the salt water – and his unshaven face pushed him from 26 to early 30s. Straight from the selection course he was drafted direct to 5 Troop SBS, M Squadron with me. The troop had been known as a loud, partying, good time and 'not a lot of work' type team, as the commanding officer (CO) put it to Rupert and me when he told us to take over and sort it out. I felt Rupert was a bit overawed: this was his first appointment within a Special Forces unit. However, I knew the CO would never have sent anyone to this team as a troop officer without being confident that he could 'eat it up'.

Rupert's confidence soon grew, and as he became more sure of himself, so I trusted him. And so did the team – despite his inability to get up a caving ladder when he first arrived. He wasn't the best swimmer in the world either! An educated, lateral-thinking man he could soar with the eagles as well as the donkeys; the troop respected him because he could speak to them as people instead of inferiors, not something you see much in the Army. Rupert's background training was from within the Royal Marines, which accounted for a lot of it.

Rupert sent his plan to London via the satellite link. If the head shed wanted any changes, they had an hour or so in which to make up their minds. Meantime we could put our feet up and savour the sights and smells of life underwater. It takes a particular personality to volunteer for submarines. Diesel fumes and machine oil, stale sweat and fresh farts make an unforgettable environment. Oh, and rotten cabbage – there always seems to be a whiff of that thrown in too, even on the air-conditioned nuclear boats. Sub crew operate in an inherently dangerous situation where a stupid mistake could entomb the lot of them on the sea bed or send their craft below its crush depth, with predictably horrible consequences. For them, the Cold War was practically indistinguishable from the real thing and they are a different breed from your average sailor. We had a great deal of respect for them, and the feeling was mutual. They dished us up our favourite meal when on board, cheese and jam butties, and gathered around to check out the Heckler and Koch hardware, all in sexy matt black.

Amid the general banter, I thought back to my first operation of this type from an O boat. We'd completed our dive and got back on board for a well-earned beer and a meal. But when we sat down, we all became aware of a truly disgusting smell coming from somewhere. Somewhere close to me. It had been a long dive and some bodily functions don't always keep to the programme – even in Special Forces. I'd taken a dump in the big toilet in the sea: undoing my wet suit and lowering the wet suit bottoms to do the business then carrying on with the dive. What I hadn't realised was that the no. 2 had floated up under my wet-suit jacket. I'd zipped myself up again with the turd inside and it was still stuck to my back when we sat down to dinner. Somehow, there was always someone ready to retell the story and always at meal times.

The tannoy sprung to life and summoned me to the operations room. I left Simon and his team talking through the mission in excruciating detail; Alan, Chris, Paul and Dave were taking it easy – especially Dave, who was crashed out on top of a Mk 24 torpedo. Rupert had got the answer from London. There weren't many changes: he had learned how to deal with the DSF (Director Special Forces) via the CO. Over the year we had established a sound relationship with Hereford and London by doing the job in a ultraprofessional manner. This was one of many appreciations that we had sent to the DSF over the last year; at last they were beginning to trust Rupert and the team. Since some of us had been doing this for fifteen years, it was about time. I reminded Rupert that the team needed an update on the plan so any changes could be initiated. This he did and the team leaders rebriefed the changes, primarily the alteration in the H hour timing from 2200 hours to 2359 hours, meaning that we would be releasing from the submarine at 2200 hours tonight. The other key change was that an SAS team would be acting as one of the emergency response (ER) units. (Some things never change – they bulldozed their way into the Falklands crisis and we've seen their zealous overconfidence time and again in environments for which they haven't trained properly – and it showed.) We hoped and prayed that come H hour our own teams would make it there to act as our back-up.

The time was now 1800 hours. Rupert briefed the team to be ready for the water for 2130 hours which was D-150 minutes.

The last supper went with the usual chaotic, yet organised way it does on a submarine. The queue for the food was as long as ever, but the team got priority and we were served first. Everyone was in jovial mood. As the guys finished their meals they moved away mentally focusing on what we were about to do. Some simply wanted a quiet spot to lie down and relax; others, like myself, liked to keep busy, checking and double-checking the equipment. In most cases the fore-end of the submarine was the quietest area and it was a good place to sit and think. I often wondered what would become of my family if something serious happened to me. I'd done all I could: I was well insured so the family would be financially secure if I didn't make it back in one piece. On the other hand, there were less dangerous ways to make a living. And everyone knew I wasn't in this for the money.

Team brief backs complete, the team filed through to the fore-end for the final preparation of their equipment. Clambering into the diving gear and securing all weapons and equipment always took an age. There were 24 of us, elbow to elbow: what a pin-up collection, girls! Two dozen of the Britain's finest getting their kit off. We did notice one naval officer who tended to lurk in the vicinity, catching an eyeful of bums and bollocks. He was treated with the sensitivity and subtlety you'd expect from the Royal Marines: ribald invitations for grotesque sex acts that would have been the best offer he'd had in years had it been serious. He could either stand there and let us take the piss, or bugger off and miss out on what was on view. He stayed.

To survive in the North Sea we wore 'woolly bears', thick thermal suits that would keep us just the right side of frozen while we sat on top of a submarine doing 15 knots underwater. On top of this came a waterproof lightweight dry bag; this enabled us to swim and climb with relative ease and comfort. Lightweight boots were next, followed by the weapons holster which carried the 9 mm Sig-Sauer pistol and four magazines with four more for the HK. The holster could also fit a diving knife, or you could secure it on your arm or leg. This equipment fits so tightly to the body that you can't bend freely until you are in the water – then it stretches. Over this comes the 'waistcoat' which carries the day and night flare, another four HK magazines (we were ready to start a small war here), a strobe light, plasticuffs (we do take prisoners), a waterproof VHF multi-channel radio, a SARBE (search and rescue beacon emergency), and individual aide-mémoire.

We carried an HKMP5A3 9 mm sub-machine gun on top of the waistcoat, attached by a weapon sling and slung either on the right or left side of the body depending on whether you were right- or left-handed. It was secured to the body through the trigger guard by a karabiner. It was fitted with a loaded magazine and a round up the spout, safety catch applied. The last items to load on were the LAR 5 and a life jacket. Now we really looked the part: the Men in Black, ready to protect the country from the scum of wherever.

At precisely 2130 hours the alarm bells rang and the submarine began to surface. Like a line of identical Special Agent 007s we picked up our masks, fins and communications helmets and squeezed through the narrow corridors of the submarine towards the operations room. We headed for the hatch that leads to the fin door, weighed down with equipment and swaddled in our woolly bears, sweating like pigs. You sensed the change in air pressure as the hatches opened topside. Situation update passed back down the line: HMS *Invincible* was on station; the deadline given by the terrorists still held, and we had the go-ahead. The weather was closing in from the north and the decision was taken to storm the rig now, while operations were still possible.

My shoulders were on fire, but I was near the end of the climb. Friendly faces met me at the top of the ladder and dragged me over the lip of the cable deck. I made the mistake of looking down at the black water surging around the rig legs so very far below. What the fuck was I doing this for at my age?

2343: I located Rupert and checked all teams had made it.

'Seventeen minutes to H hour,' I said.

'Let's move,' Rupert replied.

The team was all around the cable deck, weapons at the ready, looking up and around, ready for trouble with practised eyes. They looked mean and they knew it.

Radio check.

'All stations Whisky 1 radio check. Over.'

The teams answered in order.

'White 1 OK? Over.' (Simon's team)

'White Alfa 1 OK. Over.' (Dave's team)

'Red 1 OK? Over.' (Chris's team)

'Red Alfa 1 OK, Over.' (Alan's team)

'Black 1 OK? Over.' (Paul's team)

'All stations Whisky 1 OK. Stand by to move. Move. Out.'

Rupert changed channels to pass a message back to the CO, transmitted via the Nimrod circling unseen above.

Dry feet. We were safely on board and moving to our targets.

Up steel rung ladders to the upper decks, every dark corner and every doorway a threat: no knowing where the terrorists might be. It only needs one bad guy to wander off for a fag or a piss at the wrong moment: so many operations like this have gone wrong due to silly chance events like that. I talked quietly into the radio the whole time, counting down the time to H hour as the teams made their separate ways to the accommodation deck.

Each team's lead man carried an MP5SD, the silenced version; ready to take out any potential target prior to the final assault. However, it seemed 'all quiet on the western front'. In gopping weather, who'd think anyone could get on the rig undetected? Obviously not the terrorists. Rupert and I dropped off the winding tail at our pre-arranged point as the team continued their move to the targets. Four minutes to go . . .

As if on cue, the wind gusted harder, lashing the rig with torrential rain. What a night! Any other time, we might have noticed the cold, but not when a door suddenly opened in front of us. Those of us in the dark froze where we were; those in exposed positions slipped into the nearest shadow. It was a woman, with an American CAR-15 carbine by the look of it. She went back into the eating area, slamming the door behind her.

Time was getting on. I spoke into the mike; 'All stations: this is Whisky Alfa 1 – 2 minutes. Out.'

The teams didn't say anything in reply this close to the enemy. Instead, they acknowledged by double-clicking the pressel switches on their radios. Five double-clicks told me they were all on the net, ready for the word. My eyes were glued to my watch. The big hand crept towards midnight.

'All stations, this is Whisky 1 – one minute. Out.'

I got another five double clicks on the headphones. I squeezed Rupert's shoulder then he told the aircraft that we were in position. The second hand crawled towards the 30-second mark.

'All stations, this is Whisky 1 – thirty secs. Out.'

Clickety-click ... Now the teams were on final approach, creeping up to the doors of their target areas. This is the most dangerous time of the operation, that hideous moment when some bozo wanders out to discover the Man from Milk Tray is at the door with a machine gun instead of chockies. Everyone was counting down from 30 ... 29 ... 28 ... Ears were tuned for the slightest sound, eyes focused on the doors. Rupert and myself watched and listened. I could just make out a distant sound of helicopter blades: the ER teams were in-bound. They were listening to our radios. They'd heard the final messages, the teams indicating they were ready.

The setting could not have been more dramatic. The rig, gaunt and black with the wind howling through the structure, the whole thing feeling like it was tipping back and back, but it was only the clouds racing past above. Rain spattered across the steel decking, puddles spreading from every corner. Water cascaded down through the rig, providing a useful background noise to drown out the sound of our final approach. I pulled my respirator down from the top of my forehead and secured it over my face.

'All stations this is Whisky 1 – fifteen secs, ten secs, five secs. Stand by – Stand by – Stand by – Go! – Go! – Go!'

Showtime. Suddenly there were helicopters swarming all around the rig. I counted three. A Chinook appeared right above me with a thunderous roar. Ropes spewed from its back, whipped sideways by the rotorwash. Assault troops streamed down each rope to spring on to the slippery steel and launch themselves at their targets. Some lost their footing on the greasy surface and somersaulted across the deck.

Staccato bursts of fire echoed around the rig. Boom-boom, boom-boom: the 'double tap' signature tune of the MP5 9 mm sub-machine gun in expert hands. I sensed the *whump* of 'flash crash' stun grenades and a hint of gas through the respirator and shoved the gas mask firmly against my face to ensure a good seal. Shots rang out above our heads. The earphone in my right ear was alive with orders, directives from individuals in the teams. A few seconds from H hour and the rig was alive with hundreds of black-clad figures. They poked and prodded every nook and cranny of the structure. More helicopters arrived, pilots wrestling with the controls to keep station in the high winds: like us, they were on the very edge of operational safety. A slight error, a sudden gust and they'd be in the sea or parked on my head.

Suddenly, the radio was alive as the guys blurted out their first reports. Their excitement was evident, yet controlled. We spent many hours stressing the importance of clear communications and of the controlled reporting of X-rays (terrorists) and Yankees (hostages). It takes tremendous self-discipline to give a clear, concise report when your veins thump with adrenalin and the air is thick with gas, cordite and, quite possibly, blood. Rupert recorded the data, directed teams that needed help and co-ordinated search patterns over the rig. Months of training had paid off: no casualties reported yet.

This kind of mission takes place within the framework of UK law. You might find yourself as 'Marine B' in some future trial at the Old Bailey, and you don't want to make a prick of yourself. Everyone remembers the Gibraltar trial. Defence counsel: 'Why did you shoot him thirteen times?'

Soldier X: 'The Browning 9 mm only holds thirteen rounds in the magazine, sir.'

This remount exercise concluded with the removal of captured terrorists and rescued hostages and the post operations procedures. The rig is then handed back to civvie control. They will do the scene of crime stuff and, for some exercises, we will go all the way to testifying at mock trials. You have to watch out for all sorts of tricks: Special Forces guys and women in this case played the 'terrorists' from attached intelligence units. The latter can be very devious, hiding among the hostages only to pull a grenade or handgun when you're not looking.

Endex. The back-up team move back to the *Invincible*. We fly back to the submarine and sail on to Scotland, pick up our gear and home for tea and medals. Well, a few beers when we get back to Poole anyway.

2 The Cruel Sea – Training 1970–73

I SECRETLY WORSHIPPED the commander of the German submarine U 16 in Nicholas Monserrat's novel *The Cruel Sea* when I was a kid. I'd read it in bed at night, under the blankets with my torch, and go to sleep dreaming about firing my torpedoes at HMS *Compass Rose*. I wanted to be part of a professional fighting machine like that. By the time I was sixteen, my parents were splitting up, and I wanted out. More than that I wanted to do something different – I wanted to join up in a fighting force, and I wanted to go to sea. That meant the Royal Marines. Their advertisements were better than the Army's, anyway; the Corps looked like more of a challenge. I had no idea then, of course, just how big that challenge would turn out to be. Or that challenges can get addictive.

I was extremely physical at school on the rugby field, reaching county standard and getting through to the Kent under-sixteen trials (I was considered too small in the final stages), but I often wondered what I was doing thinking about the Marines. I was a wimp when it came to fighting, until my brother Alan got bullied by Andrew and David Johnson – the hard boys of my village. My father promptly taught my brother and me to fight. He was a carpenter, a caring man underneath, but strict, and with arms like tractors. He could dish out a beating to remember.

One day, my brother and I were coming home from school when Andrew and David appeared at the corner of the street. Alan got a punch in the stomach from David for accusations of cheek at school. His brother looked on with satisfied glee. Suddenly, the knife appeared from Andrew's pocket. He came up to me and Alan and slit every buttonhole on our smart black school jackets. Then off they went.

'What the fuck do we say to Dad?' asked Alan.

We immediately thought of the beating that we got from him for making a planned robbery on the local butcher's shop. (As far as we were concerned he deserved it – he was a boy molester and at one stage he even had my trousers around my ankles trying to touch my parts. My brother nearly suffered a far worse fate until I found out about it. However, the beating from Dad was worth every penny that we stole from the dirty bastard – we'd get him again later.)

We imagined the beating we'd get for coming home with our jackets ruined. To our surprise Dad was cool about the whole thing. His only question was why we didn't fight back. Why did we just stand there and take it?

'But, Dad, there were two of them and they had a knife.'

Dad's response was to teach us to fight back and win. I took one lesson to heart. 'Son,' he told me, 'when you start hitting someone, you hit them and keep hitting them until they don't get up.'

We had our revenge a few months later. They tried it on with us again, and this time we beat the crap out of them. We didn't stop until they got up and ran off. We were greeted like heroes in school and, when we told Dad, sitting in his big armchair, we were his heroes too.

That day changed my life. Now I knew I was capable of anything I set my mind to. I applied to join the Marines. And one day, a gigantic ginger-headed sergeant appeared at the house: six foot plus and with a great barrel chest, he squeezed into the living room. The interview went well. No fantastic qualifications required; I answered the questions confidently as he looked me straight in the eye. The sergeant left with a smile; he seemed to like the idea that I had learned a lot about life by my frequent short visits to the Borstal Correction centre: I was not another public schoolboy who expected everything in life on a plate.

Six months later I got a letter ordering me to report to Deal in Kent on 20 July 1971. I was sixteen years old, as my dad reminded me when he saw me off at the train station. He kept going through which train stations I had to change at; in a funny way, I think he was more nervous than I was. When my train pulled out of Norwich, there was a tear in my dad's eye – a first for me!

The train clattered along the Kent coast on its way from Charing Cross to Folkestone. We had just left Sandwich, the next stop was Deal. I noticed that on board there were other ordinary-looking men and they began, like me, to get a tad apprehensive. As a junior Marine, under seventeen, I was to do my initial training at Deal, then progress to the Commando Training Centre (CTCRM) at Lympstone, on the Exe estuary in Devon. The train approached Deal station. It came to a jerky halt; the doors opened; we exercised our last moments of freedom. Some carried grips, others carried suitcases; some had long hair and others, like me, had anticipated the circumstances and arrived with shortish hair. None of us spoke; some ran, some walked. Our tickets were taken. Outside were four dark-coloured trucks and there he was: our worst dream come true. The large ginger-headed recruiting sergeant. He had a sparkle in his eyes as if to say 'no mothers now, lads – you are mine!' We clambered into the trucks and off we went. Nine months of training had started.

It was everything that I had expected. Physically hard, constructive and balanced programmes designed to build up the weakest person until you could achieve the set standards – provided you had the mental strength to keep going when every muscle in your body was crying 'enough!'

I was in Northern Ireland only a year later, serving in 40 Commando at the age of seventeen. It was 1972, the year of 'Bloody Sunday' and the height of the Troubles. We were sent to Forkhill, one of the biggest terrorist black spots in South Armagh. I thought there was a terrorist behind every door, every tree, every road-junction. I was petrified for the first few months of the tour. Only the quiet, confident professionalism of the NCOs, our sergeants and especially the sergeant major, kept me together. That and the nine months' outstanding training I'd received at Lympstone – far longer and far more thorough than the army's eighteen-week infantry training programme. As time went on and I didn't get blown up, I calmed down a bit. My best chance of staying alive was to learn how to do my job to the highest standard – and learn fast. I wanted to learn; I was enthusiastic and tried to be professional at all times.

Occasionally, at one of our VCPs (vehicle check points), we'd stop these long-haired, rough-looking characters, dressed in scran-bag clothes and smelling of peat, looking for all the world like the local IRA boyos. When I asked, my mates told me, 'That's the SBS/SAS.' I could see they were in a different Marine Corps to me, almost in a different war. These enigmatic guys were so self-contained, detached. I'd heard of the Special Boat Service, but until then I hadn't understood how mysterious it was, a force operating at the extreme margins, playing cat-and-mouse with the IRA. I found myself drawn to the idea.

At the end of my first Northern Ireland tour, they sent 40 Commando Royal Marines to Malta. It was short-lived as Dom Mintoff, the Maltese Prime Minister, demanded more money for allowing Britain to retain its presence on the island. No chance, said London, so we were pulled out. Instead, they sent us to Florida for a few months, on an exchange with the US Marines; Libya moved into Malta – good riddance. This was a revelation for me. Sunshine, space, easy living: I've loved America ever since. When the brief exchange tour was over, I applied to join the Special Boat Service – the SBS.

Selection for the Special Boat Service, the maritime equivalent of the SAS, was a unique course back then. (Today, the selection course has been combined with that for the SAS – a retrograde step in my view.) Our company sergeant major told me I could do it, but that my extreme youth – I was still only nineteen – would count against me. But despite reservations, they accepted me for Selection. My physical training started on the Commando Carrier HMS *Bulwark*. I ran around the flight deck night after night once the ship had secured from flying stations. I'd be there, rain or shine, ably supported by Section Commander and good friend Jeff E. He was full of enthusiasm and professional pride for the Corps, Unit and the Troop. Jeff kept me on the straight and narrow as a young marine and kept me focused throughout the months before I departed for Poole, Dorset where the SBS headquarters and training establishment is based. We didn't meet again until eighteen years later, when I was on my advanced command course (promotional course for Warrant Officer) at CTCRM Lympstone. By this time Jeff was a regimental sergeant major (RSM), which was not surprising. We had lots to talk about!

I was the youngest ever person to be selected for SBS training. And it was hell. Of the 43 people who started with me on day one, 22 had dropped out by day fourteen. I was so physically and mentally tired that I've never been able to remember the details of what we did. It was a killer.

Many applicants resigned after the initial selection phase, concluding this was not the life for them. There were just nine of us left, six Royal Marines and some extraordinary foreign soldiers: one Norwegian and two Iranians. Relations between Britain and Iran were said to be strained. However, the SBS had a man on exchange to the Iranian Special Forces at the time so the political situation cannot have been that bad.

The two Iranian Special Forces guys were Sergeant Essi Badi, who couldn't speak a word of English and relied on his colleague, Pavis. Essi was about 5 foot 8 with typical Iranian features. Extremely fit, he was a far cry from the Arab 'special forces' types. I remember thinking that if they were all like him in Iran, no wonder neighbouring countries like Qatar and Bahrain were on the defensive. (I heard he was later assassinated after rescuing some of the Iranian Royal Family from the Islamic revolution.)

His colleague, Captain Pavis Zanardy, was a gentleman. He spoke very good English and although a totally different person from Essi, he was a tough, lateral-thinking soldier. Essi gained our respect as the course progressed, but we liked Pavis from the start. And we loved his wife. She was living proof of the legendary beauty of Persian women: an absolute stunner, great sense of humour and spoke perfect English. (Also connected to the Iranian Royal Family, Pavis and his wife were resettled in America by the CIA after the Islamic revolution.)

The Norwegian SF officer was a massive, Father Christmas of a man, Lieutenant Peter Magnussen. A jovial giant not blessed with the best physical body, he made up for it in many other ways. (My last contact with Peter was at Tromsø. He had left the SF circle, like all good officers do, and had taken a post as the hospital personnel administrator.)

Marine Zenor, later to be commissioned within the Marines, was the intellect of the course. He kept himself to himself. His parents were from Poland and Zenor could speak several languages. He maintained the pace to the end of the course until the resistance to interrogation (RTI) phase. At this stage for some unknown reason other than exhaustion, Zenor started to speak fluent Russian to the interrogator who responded with Russian. For this reason Zenor was not accepted in Special Forces.

Physical training three times a day; up at all hours ensuring that you are able to map-read in any light in any terrain; conducting limited diving aquatints to ensure that you could dive both night and day and finally ending up with a quick 30-mile cross-country trek. On completion the training team felt that they had about the correct numbers – the pre-selection was over – and now the selection proper began for the next four and half months.

I realised that it was a complex business that can easily become out of perspective, unbalanced and an end in itself if the correct attitude was not enforced from the start of the programme. The purpose must be kept in mind at all times. It intended to prepare a man for the four major types of SBS employment. These require both special training and a special man. Above all, the training had to be current. It is no good training for what was required in the past: we had to think what would be required in the future. The service always asked themselves the question, 'how relevant is it today?' and that is where our standards came from.

Notwithstanding the above, it was important to prepare a bedrock of expertise, military skills to which further requirements could be added.

The next realisation, or aim as I saw it, had to be that SBS training was both selective and constructive. To cope with the peculiar circumstances of SBS operations a man must be special or different from the ordinary (note – not necessarily 'superior'; I later met far superior soldiers in units such as the Paras, Royal Marines and 14 Int. Coy). However, what the SBS man also needs is something more than natural aptitude to cope with the increasing complexity and sophistication of some of our Special Forces skills.

It was explained to us at the beginning of the training that there were too many young men today whose bodies, upbringing and attitude had not prepared them for the standards they were looking for, but given time they could mould us into useful SBS operators. Although some form of sifting process was necessary, the aim of SBS training – in keeping with that of the whole Corps – was still to 'teach' not 'test'. This proved to be a valuable attitude towards training and the man.

It was easy to see that the standard of each man could be lifted above the level he would reach unaided in a 'take it or leave it' style of training. Every buried quality was brought to light by the instructors, whose ability to instil confidence into young inexperienced marines was truly inspirational.

When an SBS task must be carried out covertly (perhaps it has high security implications or is particularly difficult or delicate) it requires small numbers of men to work in isolated, unsupported and/or hostile conditions. Many missions involve secure movement to and from the target. Many targets are on or under the water. This is incredibly demanding, so the potential of each individual must be brought out to the fullest extent. Good training imparts confidence. I've seen a lot of guys achieve things they didn't believe themselves capable of when the task was set. This was what I was hoping for all along. This was done in three ways: firstly, by provision of motivation – and this means something more than good fun and an enjoyable way of life, although it does help! That appeal leaves too much room for alternatives.

In outline, we (the SBS) were the appeal to the volunteer. The Special Forces style of warfare was the in-style. Guys would put in a great deal of effort to get the badge. It is a difficult form of warfare to master because it is complex, seldom clear-cut, unorthodox, individual and, at times, messy and underhand. For this reason, it is disliked by many regular-minded soldiers, but is a challenge to others – men who are not necessarily superior, but different. There is an inter-regimental/international rapport and camaraderie between Special Forces operators. Special Forces soldiering requires the use of the whole dedicated man – mind, body and spirit. Most jobs require one of these only. Our service was completely absorbed and involved in its work. Its standards and performance are the concern of all, because all can affect or influence them. Every individual is important. It was stressed to us that we are the only organisation of our kind in the UK, and that there are only a few organisations exactly of our kind in the world. Most countries have frogmen but few have frogmen who are also parachutists and canoeists and who are trained for an operation in the Arctic, mountains, desert, jungle and city.

The training was by leadership and example – this inspired the mediocre or ordinary like myself to extraordinary efforts and standards. The sharing of the physical and field activity of the students by the instructors was the rule and not the exception. Instructor participation demonstrated that the distance could be covered or the conditions endured by anyone with the personal discipline and determination to succeed. Personally, I always maintained a fitness level way and above what was required. 'Skill, fitness and mental attitude' can enable you to surmount every challenge, 'in comfort and enjoyment and with the thinking mind unimpaired' as the manual said.

Training hammered home the lesson that tactical procedures must be followed instinctively. Equipment – kind, type, modifications, organisation and fitting, given thought and attention – is effective, and helps overcome the distance, condition or situation. By training, design and programming, the instructors aimed at a gradual but consistent build-up of the man. This is the overriding factor. By ensuring everything you do, or are told, contributes towards the overall aim, you understand the point of what you are doing, and are encouraged in your efforts by the thought of the end results.

This applies to all swimmer-canoeist training, but is of particular relevance to the SC3 Qualification Course. Sub specialisation is essential. Unless you are going to become a jack of all trades and master of none, the modern SBS operator needs time, facilities and expert instruction. All three of these requirements are in short supply in the SBS.

We are invariably overstretched and we are liable to become a much younger branch than in the past. Only individual and section sub-specialisation can provide the mix of skills necessary. Where possible we now use outside agencies for training – people who do have the time, facilities and depth of expertise we are looking for. For example, the individual sub-specializations of communications, demolitions and photography go for training with 22 SAS and the School of Service Intelligence.

For similar reasons it is not possible to cram into a swimmer-canoeist (SC3) course all the knowledge that we as SC3s might at some stage need. The course is cut to the minimum necessary to give the man a grounding in our main skills of boating, diving, land movement and tactics. It introduces us to the skills we will later have to master and begins a sifting/moulding process that eventually produces the man we need. Thereafter, in slow time (between six and nine months) we will conclude SC3 continuation training, which covers those aspects essential for all SC3s and his task sub-specialisation. This way there should be no risk of half-learned knowledge and unfair pressure on the student.

Because later operational employment is likely to be at the expense of training all SC3 training must be thorough and lasting in its effect. For this reason SC3s would not be absorbed into operational teams until they had the opportunity to complete their SC3 continuation training and if possible, their sub-specialisation course. Our briefing ended with the training officer saying that if we did have to be allocated to a team, this team would be reserved for training or Poole-based duties. Only under exceptional circumstances and with exceptional SC3s was it fair to employ us on operations.

SBS training began. We were given some bizarre initiative tests. One morning, at the crack of a sparrow's backside, we were driven to some woods in the middle of the

Pennines. The instructors made us strip off and took our clothes, money, our ID – even our watches and rings – leaving us in our underpants. They left us our boots, but removed the laces. We were each given a roll of hessian sacking, and told us to make clothes out of it.

'I never thought they'd give us fucking needlework,' grumbled the guy next to me as he struggled to thread his sailmaker's needle. By the time we'd finished sewing and put our sackcloth on, we looked like a band of medieval lepers on an away-day. The merry sergeant now divided us into six-man teams. Each team got a heavy four-wheeled bomb trolley, the type used to move ammunition. Each trolley had a cardboard box on top. We stared at these things in amazement.

'With your trolleys you will make your way to Portland naval base by eleven o'clock tomorrow morning. Otherwise, you're finished, Returned to Unit [RTU].'

We looked at him. He was a real joker: very Monty Python. Portland was more than 350 miles away. Dressed in rags and shoving a bomb trolley . . .

'However, before you leave, here's your dinner,' he continued. On his signal, the other instructors opened the boxes and out of each one came a chicken. The chickens had obviously been listening to the same announcement, as they bomb-burst into the woods, with us in belated pursuit.

It seemed an impossible mission to get back to Portland in time, but we'd reckoned without the Great British Public, whose coolness in the face of so much silliness was incredible. Shambling along in our sackcloth and ashes, pushing this bloody great trolley, we were amazed to find that everyone we met, almost without exception, took instant pity on us. By the time we got to Portland the next morning, my team were all in suits, mostly WWII demob Vintage, donated by kindly grannies; we could hardly move for all the food that had been shovelled down our necks. We even arrived with two hours to spare: whenever we'd needed one, a truck had stopped to give us a lift. The only difficulty we'd had was extricating ourselves from all that generous warmth and moving on to Portland. It was as much a test of social skills as of initiative.

During the initial fortnight no one told us anything about the SBS, or what the Selection process proper might hold in store. I thought it could hardly be any more exhausting, physically and mentally, than this winnowing-out process. I couldn't have been more wrong.

The difference was diving: one of the most taxing things – physically and mentally – you can ever do. And if you don't pay attention to the big picture as well as to every last little detail, it's also one of the most dangerous. If you haven't understood the principles of diving by the end of the first month and you're winging it, then you'll probably have a serious accident in the second, when you move on to a tactical set fitted with a re-breather so that it doesn't give off any tell-tale bubbles.

It's hideously complicated. You're operating at night, in a pitch-black lake choked with clinging weeds. There's no moon, you're cold and tired, you're being pushed past the limits of your endurance, and under those conditions it's extremely easy to fuck up. When you're the eighth person into the lake, and the mud's all churned up, and the visibility is so bad you can't remember which way is up or down – that's when you find out whether you're claustrophobic. And there's one golden rule to diving that should never, ever be forgotten. If anything goes seriously wrong underwater, it's always far worse than something going wrong on land. Because you're going to drown.

The training was almost continuous; we had one day off a week if we were lucky. The dangers of the water and how quickly things can turn sour in this unpredictable environment were hammered home day after day. We were supposed to complete three or four dives a day.

The dangers of working with volatile gases were drummed into us. Oxygen is the most dangerous gas that anybody can work with. Funny really. In a recent conversation with a doctor who had just completed his PADI Instructor course, he asked me what I did for a living. I explained that a lot of my time was taken up with diving. He asked what diving I did so I told him that I was qualified on Air, Oxygen and Mixture. His immediate response was, 'How long have you been diving on oxygen?'

'Twenty years,' I said.

He stood back in amazement and called me a liar: I should be dead now. I tried to explain to him that it is not the case with oxygen. With his vast experience in the diving world he walked away convinced I was bullshitting.

Our first Ship Attack was unforgettable. My partner was Gordon and, looking back, we were both extremely nervous. It was a daylight attack and we were briefed that our particular target area was the port side prop shaft of HMS *Plymouth*, an old frigate. We were to be dropped off from an LCU (landing craft utility) about 2,500 feet away. We checked and doubled-checked our equipment. In the morning we had been dropped off at the stern of HMS *Plymouth* for our first experience of being underneath a 6,000-ton ship. This was relatively easy as we were placed at the stern and swam the length of the ship. It was deathly quiet and very murky. The workings of the ship were shut down so we got a rather false impression from this exercise. This was to change on our next trip to the ship's bottom.

Gordon and I stood on the ramp of the LCU ready to enter the water. We jumped in, adjusted and checked ourselves in the water, gave the thumbs-up and took our compass bearing to *Plymouth* and swam. It was like swimming through cold soup and the noise of the harbour vibrated through our ears. Thirty feet down and it was pitch dark, like diving at night. The closer we got to the ship the noisier and darker the water became. We concentrated on a depth of 22 to 26 feet, trying to maintain a compass course towards the ship. Suddenly, the compass board touched the side of *Plymouth*. We'd made it. As briefed, we turned right and headed for her stern. It was an eerie experience to be right alongside a ship with her engines running. You feel the sound throughout your body, not just your ears, and I found it horribly disorientating.

Minutes later we saw the silhouette of the prop shaft protruding from the ship's stern. We were supposed to secure our limpet mines to where the prop shaft joined the hull. Once we'd placed the mines, we took a reciprocal compass reading and left the noise and clatter of *Plymouth* behind us. Our brief was to swim some five minutes from the target and surface and be picked up by the safety boat.

We were elated when we got back. Both of us were quite full of ourselves for completing the mission successfully. Then we reminded ourselves that this was a daytime attack, the next one was the tester: at night! Ship attacks like this were practised day and night after this.

Then it was on to submarines. Like the decision to jump out of a perfectly serviceable aircraft or dive beneath a 250,000-ton oil tanker, swimming out of a sub underwater is another mental hurdle to overcome. We started at Portsmouth and HMS *Dolphin* where the submarine escape training tank was located. The tank was a landmark in the Portsmouth area: standing 120 feet high, it dominated Gosport. This is where all submariners conducted their training prior to joining a submarine. It was slightly different for us. We would be required to train on a submarine about four times a year, so we had to retest here annually.

To be able to board a submarine you needed to do the tank training, which simulates exiting from a submarine 120 feet down. It involved medical tests followed by three 'free ascents' (an ascent without the assistance of diving apparatus) from 30 feet, 60 feet and 120 feet. This training was conducted over a period of two days and on completion we then moved to a submarine to do the SBS exit and re-entry training. This was conducted off the southern coast of England or in Scotland at Loch Long or Loch Striven. It is a means of infiltrating/exhilarating into a target area covertly.

The exit cycle consisted of leaving the submarine from the forward single escape tower. Once out of the escape tower you moved to the lurking area and waited for the remainder of the team to exit. We used the RABA (rechargeable air breathing apparatus). The submarine is rigged with 160 cu. ft bottles positioned in the fin and the lurking area. This enables you to recharge your bottles. Depending on how fast the submarine is travelling, swimmers 2, 3 and 4 exit the submarine and proceed to the lurking area. By this time number 1 and 2 swimmers have released the stores in the lurking area, which consists of the boat's engines, radios and personal stores. All this kit is sent to the surface on a line until the swimmers are together and move up the line to the surface.

Once on the surface you inflate the Zodiac and load the stores and engines. When you're happy everything is set, you cut free from the submarine and proceed to the target area of operation (AO).

Getting back aboard – the 'entry phase' – is slightly different. First, you have to establish communications with the submarine. Then tie off the boat on to the submarine's periscope. The casing diver (from the sub, in a training scenario) checks the submarine is ready and brings up the RABAs to the guys topside. Each diver swims down the periscope to the fin where they recharge their RABAs; then you go down to the casing and assemble in the lurking area. Provided the sub slows down, you can move forward and re-enter via the one-man escape chamber. By a series of taps and bells the submarine will then close the chamber and drain the water from the tower, and the lower hatch of the escape chamber is opened and the swimmer completes his re-entry into the boat.

I made one mistake in Selection, and it nearly cost me my career. The latter part of the course is spent in

Scotland: seven weeks in the Navy building at Greenock where advanced tactics were taught in the form of hard and realistic exercises in the hills of this lovely country. We did not see the Navy building that often: no sooner were we in and deservicing/reservicing equipment, we would be out again. We may have had one weekend off during those seven weeks. The weather was always wet, windy and cold. You certainly learned to look after yourself. Out on a tactical exercise, lugging Klepper canoes across those Scottish hills, there was nothing more you looked forward to at first light in the morning than a change of clothing and a cup of tea. The Scotland training phase ended with a formal test: a 30-mile open water Klepper canoe paddle from Greenock down to the end of Loch Long and back again.

Loch Long is a big sea-loch, one of the most treacherous stretches of water in the world. It holds bad luck in its black belly, and I've seen too many accidents there not to hate it. That water is really open. I know, because I thought I was going to freeze to death on it.

It was there I made my mistake: I put a life-jacket on inside out. This sounds trivial, but if I'd had a problem underwater, and I'd inflated it, the life-jacket would have exploded, and killed me. The instructor was adamant: I'd failed the entire course. Pure misery flooded through me: all this horrendous agony of effort for nothing! But my luck hadn't yet run out. Peter Magnussen insisted I had the talent to get through.

'He's young and he'll make good. Give him a chance,' he argued. 'He'll never make that mistake again.' Incredibly, my course instructor, Dick G, changed his mind.

By the end of those seven weeks, I was a different person. It had gone by in a flash, but even so I could see the subtle changes in myself: SBS Selection had pressed the fast forward button on my growing-up process. On that first day, I'd looked at the NCOs and thought they were like gods – their ability, maturity and amazing professionalism made them seem like a race apart compared to the guys on the course. I'd told myself: 'You'll never get that good.'

But now I could at least see where I wanted to be. I even had an inkling of how to get there. It felt like learning how to be one of them. And that was a great feeling. Experience has shown that the man who can best meet these demands is one who is at ease in and on the water and has a genuine affinity for wild terrain and the elements. He must be responsible, self-sufficient and resourceful; and capable, after training, of working to a directive. He must be well motivated and enthusiastic for the work, possess the self-discipline to keep himself fit and, most important, be able to get on well with others. Have we proved this was the question?

Up until that moment I'd just been another number, a nuisance kept at arm's length by the impersonal instructing staff. But suddenly there was the odd little bit of acknowledgement coming from them, the hint that I might in fact be a human being after all. Which is another great boost, when you've spent four months feeling like a useless underwater troglodyte or an overland tractor. Selection ended and to my relief all the course that finished passed. Now the real work was to begin.

3 The Silent War – Oman

I GOT MARRIED ON 3 April 1973. I'd dated Fiona for three years, so she knew the score, but we were both gutted the following day. I was ordered to fly to Oman on 6 April to do a nine-month tour of duty in what was then the no. 1 hot spot for British Special Forces. This rather set the pattern for us; our marriage has been punctuated by short notice postings, which always come up at the wrong time.

It was known as the 'Silent War'. For some reason, the operational name, 'Operation Storm', remained secret until quite recently. Our mission was to assist the Sultan of Oman's forces in their war against communist guerrillas. The British forces stationed there were known as the BATT (British Army Training Team) but the 'advice and training' often took the form of a firefight with the rebels. I have to admit that I didn't know much about the place or what was going on at the time.

The Sultan, Qaboos bin Said, was 'one of ours'. He'd entered the Royal Military Academy at Sandhurst in 1960 and graduated successfully. Qaboos went on to serve for a year in a British infantry battalion in Germany. While he immersed himself in western technology and culture, his father, an unreconstructed Arab despot, failed to deal with a growing communist guerrilla movement. Sustained from across the border in Yemen, the rebels gradually drove the government forces back to the coast and, by early 1970, they effectively controlled the southern province of Salalah, where Qaboos had been born. The British government wasn't prepared to let the place fall to the pro-Soviet rebels. The last thing we needed was a Soviet base at the mouth of the Persian Gulf, down which pass that endless parade of oil tankers on which the western economies depend.

If Qaboos's father wasn't up to the job, then his son would have to do it. Qaboos successfully launched a coup attempt in July 1970, following which he was declared Sultan.

By 1973, when I was posted there, the counter-insurgency campaign was well under way, but it was an SAS show and they wanted to keep it that way. They regarded us as needless wasters, intruding on their private war.

Our SBS team consisted of only five people. The team leader was Col who had been on my Selection course. His no. 2 was Nick, a man of immense physical strength and determination. He oozed enthusiasm and would only accept a maximum effort from all concerned. If there was one man that I wanted to mould myself on during my early days in Special Forces, it was Nick. The fourth team member was Andy, a quiet unassuming man who was always there when the sharks were around during the beach recces.

BATT was planning an amphibious assault in the Salalah area. They needed us to do the beach recce. We spent three months studying every angle of attack before Omani troops, led by an SAS team, came ashore and recaptured the town of Salalah. Then it was inland, small patrols, often a combination of local troops and British SF. Most of the SAS boys were on side, but there was a minority who didn't welcome us straying on 'their patch'.

On completion of the beach recce phase and the recapture of Salalah, we were split up and attached to the

SAS teams in the mountains. I didn't think much of this, given the undercurrent of rivalry between the two units. However, guys in both units worked at patching up a creaky relationship and it worked out much better than I'd feared. From Salalah I was flown out to 'Rocks', a position that overlooked the Yemen border. We rode to war Vietnam-style, in a Huey, the doors locked back with everyone craning their necks to spot a possible missile launch. The pilot came in at 5,000 feet, out of range of the SA-7s the guerrillas favoured. Then he spun the helicopter straight down. I gripped my harness as it plummeted to earth, slowing only at the very last moment before setting down in a cloud of dust.

My team leader was a gigantic Fijian called Lomu. He was a sergeant in the SAS and winner of the Military Medal for his bravery at Mirbat. This was an SAS position attacked by overwhelming numbers of the rebels, beaten off just before it turned into a British version of the Alamo. Lomu was the classic silent giant, very quiet, and a terrific tactical leader. He'd lost half his back at Mirbat but was still fighting. There was another SAS guy called Ging C whom I got to know really well: another quiet character who had no time for SAS/SBS rivalry.

Lomu introduced me to the rest of the team, then took me on the tour. Orientation was vital: our perimeter was guarded by a hidden shield of claymore mines, intended to greet any guerrilla raiding party with a lethal shower of ball bearings. I learned where the bunkers were and which routes to use in and out of the camp. Less encouraging was the news that the guerrillas had taken to shelling the place with long-range artillery from over the Yemeni border. But it was fascinating nonetheless and, as a young SF soldier, I couldn't have had a better leader than Lomu.

BATT was not just working with local Omani troops. The effort against the rebels brought in a big Iranian contingent and there were Pakistanis and others too. We directed a company of Pakistani troops towards a place called Raicut. They took some directing. We never got any kip because you couldn't trust them to stay awake on stag. If you went up to one of their sentry posts in the middle of the night, odds on they'd be curled up fast asleep.

Their gun groups were OK. If it came to a firefight, they'd point the machine guns the right way and get on with it. Despite their low level of training and the crap money, the Pakistani mercenaries kept going even after some horrendous casualties. Often as not, we'd find ourselves fighting past lines of Pakistani wounded, carry the enemy position, then come back for them once we'd gone firm.

I've never faced so much firepower as we met in Oman, except for one time when a company of Argentinians caught us in the Falklands. The rebels were armed to the teeth; their weapon of choice: the Kalashnikov AK-47. Built like a brick shithouse, it's the most reliable service rifle in the world. The Adoo, as the rebels were called, never seemed to clean the things. But they never had a stoppage. The Communist bloc supplied them with shed-loads of new kit. In one sweep and search operation we found 400 brand-new AKs hidden in a cave, stacked there with the ammo all ready to go.

Being Special Forces we had some latitude as to what we used. I preferred the American 5.56 mm M16 to the British 7.62 mm SLR: automatic fire and lighter ammo. But you could get into some really long-range firefights up in the mountains, so I put the bipod from a Bren gun on to my rifle. Stabilised like that, I could shoot out to 700 yards.

The problem in Oman was that the Adoo knew the terrain like the back of their hands. They were bush fit, tribesmen who'd had guns in their hands since they were children, and knew how to use them. On the other hand, so did we. And as many firefights showed, we shot even straighter. I shot a lot of rebels in Oman. More than I care to remember. In doing so, I learned how to soldier – the hard way.

They nearly always had the advantage of the high ground. Tactically, they were very switched on and, in classic guerrilla manner, they would attack only when the odds were really in their favour. If we made a major push, they'd just try to get out of the way and ambush a few patrols to keep us busy. We had air support in the shape of Hawker Hunters and BAe Strikemasters, but the Russians shipped in a load of SA-7 'Grail' shoulder-fired heat-seeking missiles. Crude things, they often tried to home in on the sun and are fairly easy to avoid by dropping flares, but they made the pilots' lives more exciting. We were in radio contact with the pilots when we called in an air strike and always tried to warn them over the radio if we saw a missile launch.

I remember one such incident vividly. The team was out patrolling, just a bit of 'hearts and minds' stuff, nothing special. The air above was constantly busy with re-sups of ammunition and food from the helicopters. Above them was the 'top cover' in the shape of Hawker Hunters or Strikemasters. Suddenly, we saw a trail of smoke shoot up from the ground some way in the distance. It curved to follow a Hunter as it turned; the missile seeker head had obviously locked on. Somewhere in that arid landscape ahead of us, the guerrillas stared up at the sky, praying for it to connect and bring down the jet. Lomu's reaction was lightning fast. He grabbed his SARBE Mk5 from his harness and flicked up the antenna.

'SAM! SAM! SAM!' he yelled. The reaction above was immediate. The pilot followed his immediate action drill: yank back the stick and go to emergency power. The Hunter stood on its tail and went up like a rocket, followed at an ever shorter distance by the missile. I don't think anyone dared breathe. Then the SAM fell away, its fuel gone or motor burned out. It fell back to earth and we all heaved a sigh of relief.

4 Wet Feet

I LEARNED THE IMPORTANCE of good communications while I was in Oman. If comms failed, our small patrols were liable to be chopped up by the Arabs before you could say 'wrong frequency'. An OP could spot all the targets it liked, but if it couldn't communicate back to headquarters, the jets stayed on the ground and the enemy went on their merry way. I loved the HF radio kit. It was a real challenge to communicate over perhaps 4,000 miles, using a little HF set to talk to our base in the UK. There are lots of specialisations within Special Forces. Patrols need to have experts in every field, but the para medics, demolitions guys or free-fall parachutists aren't required every day. Radio operators are.

This was a job that would always be in demand, but a lot of SF operators shrank from it. First off, it's a lot of responsibility. If comms go down, it's up to you to sort it out; in war, you won't get a retest. Secondly (and more importantly) you have to carry the kit, and you can end up with an awful lot of extra weight. I learned on the job in Oman, with the constant presence of a real live enemy to focus my mind. When I came home from the Middle East I kept an HF set at home and called up teams all over the world. Once, a team on an operational deployment ended up relaying urgent messages via me to Hereford. I was always on a high when working thousands of miles with this small yet powerful piece of equipment. The importance of communications was brought home to me the hard way, not only in Oman but also virtually on my own doorstep – the west coast of Northern Ireland.

Submarine and canoe operations had been rehearsed for years, new kit and new techniques improving on the impressive combat record of the Commandos during World War II. The use of canoes as raiding craft was pioneered in the summer of 1940, when British forces had retreated from France and were holed up in the UK, awaiting the German invasion. Some enterprising spirits, soon championed by Winston Churchill, planned to take the war to the enemy, regardless of the grim strategic situation. Commando raids on enemy-occupied Europe gathered pace from 1941 and have been the subject of many classic 1950s British war movies. The most famous raid, Op Frankton, involved a canoeist raid on Bordeaux, attacking German shipping with limpet mines. Six canoes started, but only two made it past the guard posts upstream to the ships. Captured commandos were executed by the Gestapo. Two men survived: Blondie Haslar and Bill Sparks, who is still alive today.

Canoes are still used by various special forces units worldwide. They were used in the 1970s by the Rhodesian SAS for cross border raids, and by the British SBS during the Falklands War in 1982. The canoe is incredibly versatile: it's light, robust and easily maintained. You can strip it down quickly and easily to produce two-man-portable loads. Canoes can be launched from surface craft or submarines, 'float offs' or 'over the side launches', or you can paddle all the way, combining movement by water with overland marches, culminating in a canoe extraction to a landfall or parent base.

In the SBS we use the Klepper Aerius II two-man collapsible canoe, which has a lightweight wooden frame with a canvas and rubber skin, and an aluminium rudder assembly. It can be packed away in three bags and is man-portable. The working speed is 1–2 knots depending on weather conditions. It can carry up to 500 lbs of stores. The advantages of using a canoe is the low silhouette that makes it difficult for enemy look-outs to spot; it is extremely quiet; it is not prone to mechanical failure, and is easily camouflaged or cached. It can be towed at up to 5 knots by a submarine or mother craft.

It has a few obvious disadvantages: it needs a minimum of two fit guys to operate, and it has a limited load capacity (if you consider 500 lbs to be limited!). It is slow in comparison to other, powered assault craft. Also, using a canoe in sea states above, say, force 4, can be pretty sporting. The canoe is vulnerable to damage during beach insertions, so you can be left high and dry on the enemy coast. Being manually powered, your own body imposes limits too. It is tiring work, even for the super-fit. And, most importantly, 'quick escape' and 'canoe' never appear in the same sentence. However dangerous the situation, you can't paddle any faster!

The use of the canoe and the packing of the necessary equipment work on the same theory as an SF soldier on the ground. We'll start with first line kit. This is kept on you at all times. Your individual weapon is secured to the craft by a strong line and karabiner; the muzzle is wedged into the paddle-housing sleeve to the no. 1's right or left, depending if he is right- or left-handed. The no. 2's rifle is kept with the butt in the paddle-housing sleeve, again on the same side that he shoots from. Moving to second line kit, your webbing can be carried externally behind you, held in place by bungy – but also attached to a secure point by clip or krab. This is usually done on short paddles or if the water is very calm. However, on longer transits, rougher water, usually in seas, your webbing has to be secured internally, held between your legs. Third line – and any additional kit – is broken down from the Bergen (backpack). Although a new Klepper arrives with a seat, we always pull this out and put in its place an empty Bergen and a roll mat. Like everything else you put in the canoe, this is secured to the frame so you don't lose it if you capsize in rough waters (as you always do).

Your kit needs to be 100 per cent waterproof when using canoes. The bags need to be of a smallish size and strong. We stow three canoe bags at the bow. After these, the contents of the Bergen can be stuffed in. You attach a security strap around the goose neck of the outer bag, a length of six feet is ideal, so that you can pull it back down once you've reached the destination. Also in this cavity go the side pouches and between the no. 1's legs (front man) is the CQR (Canadian quick release) anchor. We put a bung on the tip of the anchor, to prevent it piercing the rubber or canvas skin if it comes loose.

You sit on your Bergen or roll mat and adjust your position so the centre of gravity is as low as possible, but not so low that you catch your arms or paddles on the sides or can't see where you're going. Between the two canoeists, we like to carry a pair of 66 mm LAWs (light anti-tank weapon), not because we anticipate meeting tanks on the water, but because these single-shot, disposable rockets are a brilliant source of portable firepower and will make a mess of any small boat. For navigation, we use the P11 compass, originally mounted in Spitfire fighters during World War II. (They never throw anything away in the MoD, you know.)

Clothing has improved enormously since the days of the original Commandos. You have to have kit that will keep you warm, even when soaked through. We wear lambs-wool jumpers over a LIFA type thermal top. A windproof goes over that, with a knife, whistle, torch and compass attached to it. We also carry oyster clamps, wool hats, charts, first field dressings (FFDs) and MREs (meals ready to eat – or meals rejected by Ethiopians). If in bad weather or rough seas, a Gore-Tex jacket is OK for training, but for operational use the fabric rustles so much we prefer the old wax cotton, such as a Barbour jacket. Only by practice will you learn what to wear at different times of the year. Like skiing, you must have freedom of movement, be able to keep warm when stationary and not overheat during the transit phase.

The last item to go on is your life jacket; now the SCLJ which has been slagged off due to its bulk, but it can support a man and up to 115 lbs of kit which may be attached to him. Heaven forbid! Attached to the life jacket are a day- and night-flare, a salt-water light, a whistle and, if possible, a strobe light.

Almost twenty years ago Colour Sergeant 'Woody' Nelson devised a system whereby Klepper canoes could be towed in or out of an area by a powered craft. Today's towing system is almost identical to the original: the lengths of lines have remained the same, while the quick release knot has been replaced by a para hook, and there are male and female connectors to join one canoe to another. Any 'mother craft' may be used but the idea was based around submarine work. The original concept was for a sub to tow canoes to a release point for insertion after floating them off or launching them over the side. It is safer for the sub, which can remain submerged at periscope depth. It works in reverse during the extraction phase of a mission: the canoes lie out in formation, ready to be 'snagged' and recovered by the submarine. This was what was supposed to happen on our operation, but Bill and I never had the chance after the weather overtook us.

My first canoe operation took place in 1977. British intelligence knew the IRA smuggled weapons ashore somewhere along the west coast of Northern Ireland. But it was impossible to catch them at it. Then someone suggested a classic canoeist mission straight out of World War II. It was to be the first 'live' operation of its kind, and I, fresh out of training, would be taking part.

It was a disaster.

I've never forgotten it: for the first time I let the service down. (The second time is probably writing this book!) I was still 'green and keen', serving with 6 SBS at Poole. We'd just finished PT one Tuesday morning when Dave R called the team together. We'd received a warning order for an operation in Northern Ireland. Faces lit up. This sounded better than training. A six-man team was required: one four-man unit and a two-man team would observe two different targets. Bill asked me to be his partner because I was already learning my way around the HF communications kit; in fact, he took me in preference to a fully-trained radio operator. I jumped at the chance.

Insertion and extraction was to be by submarine. We would come ashore on the Irish coast, hide the canoes, and establish OPs (observation posts) to monitor the suspected landing sites used by the IRA. It meant lying up in a cramped 'hide', eating cold rations for at least two weeks. (We took rations for 21 days, just in case.) We'd have to take great care with our camouflage and concealment, and leave no trace that we'd ever been there. Everything we took in would have to come out with us. Well, almost everything. If we crapped all over some farmer's field, we'd get rumbled sooner or later. You have to pooh into plastic bags while your partner holds the bag (cling film is now the favoured method) and dependent on the tactical situation and location, bury it well. It's a glamorous life in SF.

We had HF/VHF/UHF communications in the shape of five different radios. The operation was so politically sensitive that we were given two HF sets in case one went wrong. Headquarters did not want to lose touch with us! With all this gear, weapons and other kit, the canoes were at full weight capacity: 500 lbs.

After an exhaustive briefing at Poole and the operation cycle complete, we headed up to Faslane to RV with the submarine. It was another O class boat, HMS *Orpheus*. The submariners couldn't believe the weight of the canoes and the amount of kit each SBS guy had to carry. We rehearsed our every move over three days of briefings and trial runs.

Then we were off on a three-day voyage to the DOP (drop off point). The weather forecast was grim, with storms predicted once we were ashore. But it was supposed to be OK while we made our way ashore. Dale, Alan, Sean and Gus would follow me and Bill.

The *Orpheus* surfaced at dead of night. We had to brace ourselves inside the torpedo compartment as she pitched and yawed in the rising sea. It had been so calm under-water. Up the steps we went, through the forward torpedo hatch, to be greeted by a truly shitty night. The sea was close to the limit of operational safety. Had it only been an exercise I suspect they might have binned it.

As the submarine wallowed in the swell, we dragged the canoes through the hatch and sorted ourselves out.

'What do you reckon, Bill?' I had to shout into the wind. He looked again at the white-capped waves, as if to stare them down.

'Marginal.' He knew he didn't have long to make up his mind. 'Let's go.'

Just how bloody 'marginal' we didn't realise until we got the canoe into the sea. In the lee of the submarine, the water was comparatively calm, but the moment we pad-dled past *Orpheus*'s bows, we caught the full fury of the sea. To my horror, I realised that I'd forgotten to secure my canoe spray skirt properly. Bill was so keen to get away and impress 'jack', we were not ready. Big mistake.

We couldn't change our minds now. In seconds we were out of sight of the sub and paddling for all we were worth. I could barely make out the land, a fuzzy black outline just that bit darker than the sky. The waves got bigger. We were battered this way and that until, still a good 450 metres from the shore, we were struck side-on by another massive wave and we capsized.

I found myself in the water. Bill and I both tried to get back to the canoe, but we were swept in different direc-tions. The canoe vanished from sight and we didn't see it again until we swam – or rather, were washed – ashore. At least we got there in one piece. We found bits of canoe all over the place: it must have smashed into some rocks.

So there we were. It was 0200 hours, we were soaked through and our weapons and kit were scattered along the shoreline. We picked our way about, gathering what we could. Every time a car came past, we ducked down to keep out of the headlights. It was freezing cold, but we couldn't stop. Bit by bit, we gathered everything we'd started with except for one '44 pattern ammo pouch. So: should we try to go on or jack it in?

We decided to keep going.

When you split up a canoe, one man carries the skin, anchor and compass, the other takes the wooden frame and paddles. Depending on the operational situation, the journey can be split, making two trips, hence lightening the loads, but if you're big and strong like me, you secure your half on top of the Bergen, strap it down tightly, and make damn sure it's balanced. Although both loads are around 56 lbs each you'll usually find the stringer bag is the heavier unless there is water trapped inside the folded skin. If one load is heavier than the other they should be swapped at intervals, unless of course your canoeing partner is mega weak and skinny and can't handle steep hills! Lastly, the P11 compass should be wrapped in a roll mat to protect it from getting knocked. Thankfully all we needed to do was bag the canoe and transit the bags to a 'cache' site some 600 metres away.

If we'd known then that it would rain every day for the next fortnight and more, perhaps we wouldn't have bothered. Our kit was saturated, which made it feel twice the weight. It probably was. Hoisting our sodden Bergens on our backs, we crept inland. My boots sank into the peat soil, sometimes right past their tops and I had to wrench my legs free of the oozing mud. The ground sloped upwards at a steep angle, and we had to go forward on our hands and knees part of the way. I grabbed at tufts of grass, big lumps of mud, stumpy trees, anything to stop sliding down and having to start over again. God help us if the IRA hit us now. It would have been hopeless for us both; uphill, 130 lbs on our backs; pissing with rain; mud up to our armpits and climbing a 1:8 incline. Not a chance of returning fire.

At last we could make out the summit. There was a hint of dawn in the sky; it was time to seek shelter. We needed a good LUP (lying-up position) to rest, attempt to dry our kit and, above all, inform our headquarters what was going on. Our canoe was in bits, most of them stuffed into the rocks and hopefully out of sight. Unless we learned how to walk on water, we couldn't get back to the sub.

It had been a damn long night and frankly all I wanted was to sleep and eat. We got to the crest. Bummer! We'd conquered the mountain only to find a peat bog in front of us, stretching on for miles. Not a rock, not a tree, not a fold in the ground to hide in. Flat as a billiard table. We consoled ourselves that the area was so desolate that there was little chance of anyone wandering through it. Famous last words.

We decided to hide in the only cover available. A stream bed. It gave us cover from view and cover from the air. The only problem was that we had a very restricted field of view ourselves. We'd pretty much have to depend on our hearing, which was far from ideal. Bill sat one side of the stream and I on the other. The rain didn't let up. Daylight, of a kind, found two very wet, very pissed-off Marines hunkered down, Bill listening while I set up the radio. God, I wanted a brew!

The small 'bomb proof' gas cooker failed to light no matter how much we swore at it. Never mind, back to the hexamine block we used in training. No one was going to smell its fumes out in this godforsaken wilderness. And they didn't because the blocks were too wet to ignite.

Not a good day. The water in the stream bed began to rise. We chewed on cold rations during day 2, dreaming of a hot cup of tea and trying to forget this was potentially a 21-day mission.

Day 3 and we were supposed to be on our way to our target area of operation (AO). Was that any better? Was it hell! The forward slope was like a waterfall. We found a gully from where we could observe the target and got ourselves settled in there before first light. By dawn the rain turned to sleet and the wind got up. We were supposed to lie there, out of sight, monitoring any unusual activity in the little town. The water in the gully rose alarmingly until we were basically sitting in a river. It was pointless trying to keep anything dry. Everything was wet from day one. All we could do was wring out our sleeping bags and personal equipment every night. We had accepted that we were going to be wet and cold and that hot food or drink would have to wait.

One thing we'd been taught in training was to buddy up, putting your feet under your oppo's armpits (the warmest part of the body). All a bit girly I'd thought and it had never occurred to me that I'd actually find myself having to do it. As the days passed we knew we were getting into serious trouble. Hypothermia and trench foot were both serious possibilities. You can't stay soaking wet and freezing cold for days on end without having a problem.

By day 8 we decided to move about during the night, just around the OP. We had to keep our circulation going. Great idea, but we'd left it too late. I couldn't get any feeling back in my feet. Now I panicked. We were in the middle of bandit country and I couldn't walk! Whether it was me or the set I don't know, but at about this time we started getting comms problems. Our HQ had moved up to Faslane and 300 miles wasn't far for an HF set.

After 17 days of this we were in a sorry state. And we still hadn't seen anything worth reporting. The opposition didn't want to poke its nose outdoors either. At last, we got the word: move to the alternative pick-up point.

I clumped along with no sensation in my feet. I had to watch them to make sure I put them on the ground properly. I was carrying a lot of weight and if I lost my footing I'd probably break an ankle – at best.

We hadn't seen a soul for two weeks but now we did. We blundered straight past a shepherd, fast asleep at the edge of a field, guarding the sheep, armalites or whatever he had in there. He stirred. What do we do, shoot him? Hope that he would think he was just dreaming. We looked back. He'd folded his arms and gone back to sleep.

We eventually arrived at the cache area and checked the vicinity. All quiet. No sign that anyone else had been there. Our alternative pick-up was some 10 miles south. The covert intelligence team in Northern Ireland (14 Int.) would extract us at a pre-arranged RV at 0200 hours. If we could get there.

I got some feeling back, but soon wished I hadn't. Every step was agony. Each jolt felt like a dozen cigarettes being stubbed out on the soles of my feet. Every now and then a car would approach and we'd have to take cover.

We got there early. I gritted my teeth and clumped around to see if anyone else was about. No one. It was still ten minutes before the 14 Int. team were due, when the car pulled up and doors opened. We dumped our Bergens and leaped inside. A guy came out and heaved our Bergens into the boot. I didn't even notice, but it was our own CO. We sped to the cache site and double-checked we hadn't left anything behind. Then the pain really got to me.

It was my first experience of 14 Int. Coy. I was impressed with their professional procedures, their slickness and awareness on the ground. They were outstanding – and it wasn't long after I'd recovered that I started to think of attempting their infamously tough training course. Meantime, inside that snug, warm car, my feet began to defrost. It felt like I'd put them into boiling oil. If someone had chopped them off at that moment, I'd have been very grateful. It occurred to me that that's exactly what someone might end up doing.

It took the medics some time to cut my boots off me. I thought the feeling had returned completely, but as he prodded and probed I realised that the skin surface was completely numb. He stuck a 4-inch needle in several times and if I hadn't have been watching, I wouldn't have known about it. I was transferred to hospital at Faslane.

Five days later I was back on my feet, thanks to a team that's had more experience of dealing with these situations than most. It was about as severe a case of immersion foot as you can have, without recourse to drastic surgery. We'd found no sign of the gun-running operation, but I'd learned once again that the superb physical fitness demanded of us was absolutely necessary. That, and the mental determination to keep going no matter what. Once again, being at the sharp end, there were still valuable lessons to learn, whether it was in the heat of Oman or the oozing, rain-beaten, muddy slopes of Northern Ireland.

5 The Northern Flank

THEY SENT ME FROM THE BLISTERING HEAT of Oman to the icy wastes of Norway. For five years I was part of the NATO force deployed to defend the northern flank against a possible Soviet invasion. Had it come to World War III between NATO and the Warsaw Pact, Norway would have been crucial. There would have been another Battle of the Atlantic, this time with Russian submarines doing what my heroes in U 16 did to the *Compass Rose*. If the Russians could take the Norwegian airbases, they could support their subs far out into the Atlantic – and hammer our naval forces that tried to stop them. To keep Norway on side, we deployed a large proportion of the Royal Marines, with army regiments rotated in and out on two-year tours.

The Arctic is the most challenging military environment on earth – and I loved it. We'd spend five months on skis, with a 90-pound Bergen on our backs and pulling a 'pulk' (sledge) behind us containing our diving kit. Sometimes, we'd be in an OP for thirty days, watching, reporting on the burst HF communications directly to Hereford. We would conduct and exercise our exit and re-entry procedure from a submarine. Hereford were not allowed to conduct these procedures with Royal Navy subs but, in typical style, they decided to bulldoze their way into the area. They persuaded the Norwegian navy to let them conduct E&RE drills on their submarines – via the torpedo hatches! They were welcome to it: basically it's unworkable.

From some of the OPs you could see the Russian tank parks on the other side, their soldiers, supply dumps, the whole thing. It was weird watching them, knowing their exercises were rehearsals for overrunning our positions; and somewhere, unseen, their own Special Forces, the Spetsnaz, were lurking in the snow. We later met them face to face in Bosnia, although on a much more friendly basis. I met a tall, powerful-looking man who spoke perfect English. Was my Russian as good? In a word, no!

We were pre-planning our battle too. We recce'd every conceivable OP site that looked promising. No one liked to say it, but we knew this would be a suicide mission if it came to a real war. If the Red Hordes really did punch over the border, they wouldn't be stopped straight away (if indeed they were stopped at all). Even if NATO defences held them lower down in Norway, our patrols in the high northern latitudes would be left to fend for themselves. Someone had to stay behind to report on what was coming over the border, and maybe provide targeting information for air strikes.

I discovered that every military task took at least a third as long again in the Arctic: whether it be skiing, diving, canoeing, parachuting or just plain soldiering. There was a completely new set of standard operating procedures (SOPs) for us to learn. We had some memorable times in Norway. The Norwegian man is a stout and proud feller who has little appetite for any type of aggression, let alone standing up against a superpower. We had many good social nights in places such as Harstad, Narvik, Ramsand and a little town called Ladingan. Many times we were in a situation where we were 'over-served' but always came back for more! Some of our guys spent so long in Norway they maintained not just two girlfriends, but two families – one in the UK and the other in Norway.

As for me, I nearly ended up staying there permanently. Not warm and snug with some Nordic goddess, but frozen under the ice.

The exercise began at HMS *Faslane* in the north of Scotland. We jammed ourselves and our kit into an O-class submarine bound for Norway. It took four days to get to the drop-off point. The submarine serviced 30 miles off the north-west coast of Norway, opposite Malagen Fjord, the longest and biggest fjord in Norway. The submariners hauled our canoes out of the forward torpedo loading hatch.

Even in the pitch black of night, the white tops were visible. Each wave flecked with foam as the wind sliced into our clothing. Mad Dog, our team leader (a personality that was 'barking mad' in the nicest possible way!), paused as he stood on the casing. Then he decided it was a 'go'. Conditions were marginal, but OK. The canoes were bundled to the centre of the subs casing with enough stores stashed inside to last our patrol for 30 days. Then we scrambled into the cockpits of the canoes and gave the signal of 'red alfa' in Morse code with a torch to the conning tower. In the darkness we just saw the tower duty personnel disappear. The next thing that we heard was the sudden blowing of the submarine's Q tanks. The 'float off' off the sub's casing was in progress. The black cigar slowly sank beneath the waves. The waves crashed over the bows of the submarine, picked the two canoes up, and we paddled into the darkness. We heard a steady hiss behind us as the submarine vented the air from its ballast tanks and slid beneath the waves. They'd be back home in a few days while we did our sneaky-beaky stuff among the ice floes.

On we went, paddling on the allocated compass bearing that would take us into Malagen Fjord. But the sea was getting up now, solid waves pounding into my face and chest. Just in time, we entered the calmer waters of the fjord. The stars were blotted out above us, black mountain peaks towering over the water. It was two hours before first light and we needed a break. We'd only made fifteen miles, about half the distance we'd have covered in normal conditions.

We reached the shore and hid the canoes before selecting a lying-up position for the night. Burst transmission radios kept us in touch with Hereford. Mere survival is an achievement in these conditions, but we were going to make our way, unseen, up the fjord. Our mission was to plant mines on 'enemy' warships (Royal Navy frigates) lying at its head.

The following night we retrieved our canoes and packed all our ancillary equipment inside. The weather began to warm a little, so we looked forward to the next stage of the mission. We put off from the shore. No moon, no artificial light. It was difficult to keep Mad Dog's boat in sight as we paddled along. Only the swish of the blades in the still water betrayed our presence. At its widest point the fjord is eleven miles across and the surrounding cliffs are absolutely sheer until you get further up. Eventually a rough shoreline appears, strewn with boulders. This dramatic scene is home to a number of small villages, scattered at intervals up the fjord.

We had left our last LUP at about 1915 hours. It was 2145 by my watch when we hit the ice. Not a major drama. The canoes were designed to withstand thin ice so we crunched our way through the first thin sheet and pressed on. We'd all done this before: it wasn't that we were doing something wrong. But what none of us realised was that this complex environment had a new trick to play on us. A lethal one.

The danger wasn't the cold, but the unseasonable mildness. The air temperature was actually rising, from −20 degrees centigrade to −2. Fresh water on the fjord had frozen to create a very thin layer of sharp glass-type ice. Once through, we carried on, noting the incredible stillness of the water, like the proverbial millpond. It was so still that I noticed our bow was distinctly lower in the water than that of Bill and Mad Dog's boat. Brian and I realised our legs were very cold and the boat, now we remarked on it, was sluggish, especially in a turn.

We stopped for a navigation check and quick rest. That did it. The instant our forward movement ceased, the canoe filled with freezing water. Up to my ribcage it came and the nose of the canoe vanished from sight. What the hell? Bill and Mad Dog saw what had happened and brought their canoe smartly alongside. They stopped us capsizing.

Bill checked the front of our canoe. The ice had sliced into the lower front of the hull on both the right and left sides. Irreparable, in this particular situation. Brian and I were up to our chests as the canoe sank from sight; only the inflation tubes on either side maintained some flotation. We were in the middle of Malagen Fjord, at least six or seven miles from shore, and we were sinking.

Forget the exercise – if we didn't sort this out fast, we'd be dead. We'd thought out contingency plans for most emergencies, including this one, but we couldn't summon help as we'd intended. By sod's law the HF radio was in my Bergen, now submerged at the back of the canoe. OK, we had a back-up: SARBE Mk5 safety comms. Mine was in my chest harness and I switched it to the emergency frequency of 244.5 Mhz. Any aircraft in the vicinity, military or civilian, would pick this up. We hoped they'd hurry as we couldn't sit here too long: the bitter cold was made infinitely worse by the wind chill. Our text books gave you only 20 minutes before hypothermia would overcome you.

An hour passed. Then another. We heard helicopters in the distance, got excited, then puzzled, then bloody despondent. They flew on regardless, presumably unable to pick up our signal. Midnight. The cold was so intense it felt solid, like a weight crushing your head, the pain from my hands and feet replaced by an ominous numbness. We daren't move about to help our circulation either: any sudden movement would tip us into the fjord.

We fired off a mini-flare, then another. Nothing. We shone our torches. Every time we saw or heard an aircraft we flashed SOS on the torches and let fly with a flare. One Lynx helicopter flew right over us without noticing. By 0300 we were in real trouble. My face was frozen solid and I gave up brushing the ice off my lips. All we could do was crouch over the second canoe and support ourselves, trying to hide from the wind.

The wind drove us slowly but surely into the northern shoreline. Suddenly the canoes stopped. We were wedged into a field of pack ice that had formed during the night. We were shunted further into the ice. Then we stopped. I estimated we were about 300 yards from the shore, which was better than being out in the open water, but it was still dark, so we were still unable to judge the true distance.

Bill and Mad Dog's canoe began to sink too. Before long, both capsized under the ice. I found myself upside down in the water, unable to get out of the canoe. I came up momentarily, but spun around and my head plunged back into the freezing water. My ankle was stuck, presumably pinned by some equipment that had shifted. I had no feeling in my legs by this time, and neither did the other guys. No one could help me. I came up for air, gulped, and tipped over again. I knew I couldn't take many more duckings before my whole system gave up the ghost. Fuck! I gave it max and to my relief, I sensed my legs separate from the canoe. To be honest, it felt like they had separated from me. I was that numb.

Brian extended his hand and hauled me on top of the canoe. I sat there and laughed; the cold was suddenly meaningless, I was alive for a few more minutes. I looked around: Bill, Mad Dog and Brian were on their last legs too. The sun was now rising. If we were alive at first light we could see how far we were off shore. It was 0550 and we had been in the water for over eight hours. People will never ever believe this ordeal, and why should they? According to the experts and the books, 'the human body will survive in arctic water conditions for twenty minutes'. I tried to keep chatting and laughing, but it was so hard. Then Mad Dog decided to go it alone. 'Fuck it,' he exclaimed. 'This is my life. I'm going to inflate my life jacket and swim through the pack ice and to the shore.'

'Mad Dog,' I shouted, 'no sooner you get off the canoe, that glass-type ice will pierce your lifejacket and you're dead. Get back on the fucking canoe and do not be so fucking stupid.' He looked up in embarrassment and remained where he was.

0615 hours. I felt in my cold bones that we had about thirty minutes left before we'd die, one by one. The pack ice extended all the way and was about six feet deep. No way any of us could cross it. I shoved my numb hands into my chest harness and fumbled out our last smoke grenade. I moved to the middle of the canoe to get as high as possible. The canoes were below the surface so I just followed the line of the canoe and shoved the ice away. I pulled the pin from the grenade and held it as high as I possibly could. My hands were so cold that the smoke shooting from the base of the 83 grenade into the palm of my hand had no effect. I watched the skin burn but I didn't feel a thing.

The grenade was our last survival aid. We waited and waited. We could just make out a house on the foreshore and, to my amazement, saw a car pull away: our first sign of life in some eight and a half hours. We waved and shouted, praying that they would see us, but the car drove off into the distance – away from our screams of help. We slumped on the canoes motionless. That was it.

Then I thought I heard the faint noise of engines. I looked up and saw two Rigid Raiders heading straight for us. They stopped at the edge of the ice and the guys just gawped, jaws hanging down to their ankles. Only later did it dawn on me what we looked like: four guys sitting in the middle of the ice with no canoes, no visible means of support.

'Well, are you staying there all fucking day?' I ventured.

'Yes, mate, we can't bring our boats in there,' was the unwelcome reply.

'Bullshit!' I shouted. 'Get those fucking boats in here now! We haven't spent eight hours here for fuck all. Now do it!'

He moved the boats in towards us, still not believing what he saw. They dragged us into the boats and placed us in the centre. We could not move but it was so nice to be out of the wind at last. The boats started to move away. 'Wait!' I shouted. 'Where are the canoes – they need to be recovered as well.'

'What fucking canoes?' he said.

We pointed out our canoes just below the surface with all our stores and personal equipment in them. The coxswains didn't believe us until they saw them with their own eyes and hauled 500 lbs of dead weight on to the raiders. We then moved to the 3 Brigade Medical Squadron to be defrosted.

I couldn't walk. I couldn't even lift my legs or move my hands. I had no feeling in my shoulders and none from my hips down. The medical squadron team manhandled us into vehicles and off to the medical centre. At last we were out of the biting wind. In the hospital, doctors and nurses carried us to individual beds then stripped us of all our clothing. Now we've all had fantasies of being stripped off by gorgeous nurses (and they were too) but I'm afraid my undercarriage was fully retracted after its long immersion in the ice. I couldn't raise so much as a smile.

They put us in a big bath of hot water. It must have been a funny sight: the four of us straddled behind each other like peas in a pod. We started laughing, hysterically happy just to be alive. We remained in the bath for a couple of hours as the nurses constantly topped up with more hot water. Then we were transferred to our beds in a ward together where we were given hot soup, drinks and a meal. We remained in the medical centre for two days. The doctors could not believe that we had survived such horrendous conditions. By all rights we should be dead. It was physically impossible – but it happened. Brian and I sat in bed at Medical Squadron and we were both thinking the same: at least we were insured so our immediate families would have had their houses paid for.

We found out later that if it hadn't been for a little girl looking out of her bedroom window and shouting, 'Look, Mummy, there are people sitting on the ice!' rescue would have come too late. Her mother had looked out in disbelief, and then promptly drove to the Medical Squadron. This was the car that we saw drive away, thinking it was ignoring us. This little girl saved our lives – for which we will never be able to thank her enough. They whisked us from the medical centre back to Harstad, so we weren't able to thank her in person. The only thing we could do at that time was to send her and her family a gigantic bunch of flowers.

Looking back over those 22 years within Special Forces, that was probably the closest that I came to death. As I held that last smoke grenade as high as I could physically stretch I knew in my mind, looking around at the team, that this was our last and only chance of surviving. Once the smoke grenade had finished, I looked again at my charred hand and really thought we were done for. We discussed it with weird detachment at the time, like it was happening to someone else or was a classroom exercise. Had we done everything we could to save ourselves? Was there any trick we'd missed? None of us could think of anything. It amazed us all that every form of SOS signal had been ignored by shipping and aircraft within the immediate area.

This disaster was the subject of many conversations afterwards, including formal presentations on the training incident. Senior officers looked on in disbelief: how could these men have survived for up to eight hours in the sub-zero waters in the middle of the winter in Norway? My final sentence during my presentation was, 'As Maurice Chevalier sang, thank Heaven for little girls.'

6 Dive to Death

NORMALLY, IT GOES LIKE CLOCKWORK. This time, two people were going to die. And all because of the rain.

It poured down all through the day before and all through the night. As we rigged the submarine on Loch Long that morning, the drizzle never let up. It sounded like background radio static: a low hiss as the fine droplets struck the metal deck. We could barely see beyond the sub; its grey hull blended with grey water and blurred into a grey sky. Scotch mist if you like that sort of thing. What a gopping day to be messing about in boats.

We were practising the early theory of the SBS's methods of releasing divers from within the hull of or from the deck of a submerged submarine on to a possible MCT (maritime counter-terrorism) target. I say 'early' because this training was the foundation of these methods. The difference between these methods was simple. The former was done from within the submarine; the latter was a system adapted specifically for MCT operations, which meant that the swimmers were released from the outer casing of the submarine. During this particular exercise, we were practising the former. The MCT team exited from the submarine with their full MCT equipment. The submarine would then be dived, although it would only be travelling at a very low speed. The first swimmer to exit the submarine would be the casing diver. His prime responsibility was to check all the pre-positioned 150 cu. ft compressed air bottles inside the forward torpedo-loading hatch, accommodation hatch and the fin. Once their serviceability was assured, all was ready for the first four-man assault team. Each team exited the submarine in full diving gear, going out one by one via the single-man escape tower. The divers assembled in the lurking area and sat there as the submarine, still dived, transported them to the swimmer release point. Once on station the casing diver released our stores on a line, and we swam up the line after them. Once on the surface, we inflated our Gemini assault boats which filled out automatically once you pulled the toggle. The engines are kept in waterproof Ellistone bags; once in the inflated Gemini, they are quickly opened and the engines fitted. And off we went.

I was in the second team. The first team got the 'go!' signal and exited the escape hatch wearing their small RABA air bottles, which allowed them to move around the casing relatively freely. Once there, they plugged themselves into the big pre-positioned 150 cu. ft bottles. The casing diver watched the whole time, for safety reasons. At this depth, the RABA bottles only give less than ten minutes' air. (Air consumption increases proportionately with depth.) The sub was moving along steadily. In Loch Long we still had another 200 feet of water below the hull.

There was only one problem. At slow speed, in a mixture of fresh and salt water, the O-class submarine can handle very unpredictably.

The rain was still pouring off the hillsides, sending great gouts of fresh water out into the mostly sea-water loch. This meant that the density of the water was very variable: salt and fresh water don't mix that quickly and you can get alternating layers of buoyant sea water and not-so buoyant fresh water.

Suddenly, the O-boat hit one of these concentrations of fresh water. We lost all buoyancy and all control. The sub put its nose down and plummeted to the bottom. It felt like an aircraft hitting severe turbulence. One moment we were standing up, ready to follow the other team out of the sub, the next, we were thrown all over the place. We grabbed at anything to stay on our feet.

But it was far worse for our guys outside: the casing diver and the assault team. The assaulters were fully kitted-up with a great weight of diving equipment and weapons. And they were in dead trouble. The deeper the boat went, the more negatively buoyant they became. They were being dragged to their deaths by the weight on their backs and the steeply diving submarine. To make things worse, the deeper we went, the faster their air would be exhausted.

Inside the conning tower, the captain and crew wrestled to regain control. But they did not blow 'full Qs' – the main tanks that ran the length of the boat. If they had blown all main ballast, the submarine would have popped to the surface like a cork and the divers would have been recovered. But in 1977, the rules said that if the boat started diving out of control in a tactical situation, the team in the lurking area had to stay with it, no matter what, until the submarine's captain had regained control. Now, the rules have been changed. But it was too late for the guys in Loch Long.

The submarine sank deeper and deeper.

'When are we going to come up?' I said to Spence, who was clinging for dear life next to me. We hoped the guys topside would realise how deep we'd gone and let go. The casing diver would have a depth gauge. I knew it would be pitch black outside the submarine, but he'd still be able to see the dial and watch as the needle spun round to the point of no return. The casing diver was on secondment from the Dutch equivalent of the SBS and we prayed he would take the initiative and ignore the standing orders.

He let go at 120 feet. But he didn't tell the other guys how deep they were. He escaped alone, inflating his life jacket which was of little use this far under; by breathing all the way out as the air in his lungs expanded, he made it to the surface with seconds to spare.

Moments later, two of the remaining divers decided to get off the sub and make for the surface. Andy and Dave unplugged themselves from the big air bottles on the sub's deck. They parted company with the sub, but found to their horror that they didn't start to rise. They were at about 170 feet and the pressure was so great that the best they could manage was to stop sinking any further. Death hovered closer.

They struck out with their fins, kicking for their lives. This is where the Special Forces obsession with fitness paid off. Only the truly superfit could have got out of this situation. Andy and Dave went for it with everything they had, and just as the air in their RABAs ran out, they reached break-out depth: they became positively buoyant and ascended faster and faster. But all silver linings have a cloud: everyone knows what happens to divers who come up too fast. Despite the extreme danger, they followed their drills. They relaxed their bodies, stopping kicking and just giving the occasional flick with the fins. They came up no faster than the bubbles leaving their mouths and made it to the surface of Loch Long without a fatal attack of the bends.

Over 180 feet below, two divers were still in the lurking area. Both were married; the corporal had twin baby boys, the officer a little boy. They cracked their suit air valves and tried to go for the top. But they'd left it too late. The compressed air in the emergency bottles rushed out into their suits, partially inflating them, but it was nowhere near enough to counteract their negative buoyancy.

As it crashed ever closer to the bottom of the loch, the atmosphere inside the submarine became cruelly calm. Sixty-five men willed the captain to order the boat to

surface. 180 feet . . . 190 . . . 200 . . . only then did he snap,
'Blow full Qs.'

Compressed air gushed into the tanks along the sub's
hull. As it expelled the sea-water ballast the O-boat shot
back up to the surface.

For the men who'd just left the lurking area, his action
was now irrelevant. Cold, dank and deadly, Loch Long has
always had a sinister reputation. Because of the nearby
nuclear submarine base at Holy Loch, the British and the
Americans know every inch of the bottom of Loch Long.
Every bump and rock is charted and surveyed, and if
there's the slightest blip on the bottom-scanning sonar
trace, the operator will see it. Yet despite the massive
search that was immediately launched, using every naval
vessel and helicopter in the vicinity, there was no sign of
the two divers. I joined the teams scouring the shoreline,
but there was no trace of them there.

It took the minesweepers, fitted with a high-definition
bottom-scanning sonar, two days to find our mates.

7 Deep, Deep Undercover

MOST SPECIAL FORCES GUYS think it's the most dangerous job in Northern Ireland. And, to be honest, I wasn't sure I could do it. Even in terms of training, which was notoriously hard, I thought 'the Det' a challenge too far. In 14 Int. (14 Intelligence Company) I would literally have to rub shoulders with the IRA, drink in the same pubs, watch their patterns of activity and compile a picture of their contacts and movements. If the intelligence picture was good enough, we could catch them on their way to commit a murder, rather than after the event.

The Special Boat Service had its own demands, but I felt I could cope with them. 14 Int., on the other hand, is one of the biggest steps you can take on the SF ladder. I didn't want to stumble on it. At the same time, it was like the ultimate challenge. I had had a taste of their professional and enthusiastic attitude during one Ireland operation. I couldn't stop thinking about it. I talked it over with a few mates, and in the end it was one of them, Gordon, who convinced me I should go for it. I don't think Fiona was quite so keen.

I applied for it and, to my surprise, I was selected first time around. It was incredibly secretive. It was impressed upon me that I had to follow the instructions on my 'joining routine' to the letter. This we did. I say 'we' as Andy, who happened to be one of the operators that survived the submarine accident, was on the same course. He was a very quiet, unassuming, fit Scotsman whose professional reputation preceded him. A gentleman with nothing to prove to anyone except himself. We moved to what was known as Camp 1 (Location Unknown) for a three-week initial selection phase. Five months of Det training for covert intelligence gathering in Northern Ireland lay ahead.

The first phase of the training was conducted by the regiment and operatives from over the water. It was nice to have that happy medium of operatives actually conducting the job, instead of stay-at-homes who thought they were capable of anything – which was certainly not the case. By now, the SAS–SBS antagonism was coming to a head, simply because a unit had been formed that gathered the cream of the British services and in my opinion were becoming a threat. I found that most of the SAS thought they were too good to be taught anything by the Det; a fair number of them were failed during training. The regiment had to stop asking for volunteers to go for 14 Int. Selection, and start picking them from within the regiment. This was just before the regiment revised their attitude towards not only 14 Int. itself, but towards their own men as operatives over the water. Up until then there was no requirement for SAS personnel to attend Camps 1 and 2. This changed, mainly due to their inconsistent and unprofessional approach to intelligence work in support of 14 Int. Camp 1 was physical with a few sleepless nights, but nothing radical and beyond fit young men. (At this time women were not selected for special training.)

Andy and I relished the challenge. We were amongst the top soldiers from all parts of the British Army, RAF, Navy and Royal Marines who all had their individual goals. Most of these people did not care much for the SAS or SBS because they themselves knew their own limitations or what they wanted in life – the regiment or Poole was not on their agenda. 14 Int. demanded a lot of you, much more than pure physical fitness. Self-reliance was the top priority: you had to prove you were capable of standing on your own two feet, and make decisions in a hostile environment. Camp 1 sifted out the non-starters – and we went down from 150 guys to 60 for Camp 2.

Camp 2 taught basic soldiering and the techniques used for vehicle recognition and photography. It was wet, cold and miserable, just like Camp 1. It started the process of educating the 'squaddie' about not being a 'squaddie'. How to achieve that 'grey man' look and not look like a soldier in civvies. You wouldn't last long in Newry, wearing North Cape fibre-pile jacket, jeans and desert boots! That's an exaggeration, of course, but it was the little things that could give you away: the wrong emblem on a T-shirt, a certain ring or watch strap. The way you walk is crucial too: from the moment you join the Army you're told to keep your shoulders back and head up. We had to relearn how to walk like normal people!

They sifted us down every day. Most mornings we realised someone was missing; sometimes five or six students would have departed during the night. I felt sure I'd be next. In a way, this worked for me, it kept me on a constant state of alert. I never got overconfident: people were always watching! At Camp 2 we were taught the art of day and night photography, right through to the development of the films. I took part in a number of night exercises to hone these specialist techniques. Many nights spring to mind but one in particular has stayed with me ever since.

We had moved into a nearby market town for this particular night exercise. It was May, and our target area was on the cliffs, overlooking the harbour: a favourite spot with courting couples. We were in the hedgerow awaiting our first target: a young couple who obviously thought they were alone. They parked their car in a quiet spot and pretty soon the windows were steamed up. Andy and I crept closer, using night vision goggles. We got right up to the glass to see this guy pumping away for all he was worth. The rule was, 'one minute on, one minute off'! We ensured that the camera was set and the motor drive was functional and proceeded with the photographic exercise.

The girl was stark naked. She locked her legs right around her boyfriend and tried to put her left foot against the back window. Whether she saw the red light on the camera or our noses against the glass, I don't know, but she let out a fearful scream. We legged it. No one spotted us and the pictures came out nicely.

On completion of Camp 2 the first port of call was the nearest Oxfam shop, the Special Forces outfitter of our choice. Then it was off to Camp 3. As on most courses, you feel at a particular stage that things are going well and as long as you keep your nose clean you would be OK. This was the case with Andy and myself; we started to enjoy that confident feeling that now whatever they gave us we would do and do well – and we started looking the part thanks to Oxfam. We learned foot and vehicle surveillance and continuous recognition, covert communications, photography and CQB (close quarter battle) techniques. We learned how to recognise the make of cars from just seeing the rear light cluster or the front light cluster; descriptions of individuals: their facial features, eyes, noses, foreheads, hair, build, walking gait and dress. We learned the art of covert comms, how to fit radios within your clothing, and the radio procedures we'd use over the water. We continued with more photography and target recce reports.

I particularly enjoyed the driving training. I had a few points on my driving licence for speeding, but now I could ignore the speed limit – I had to! It was tremendous fun to bomb along, with a driving expert sitting next to you, egging you on to treat the next car as a target. You had to get past it safely, whatever the road conditions.

Whenever we went out on the ground there were a number of items that would always be with us in what we called a car bag. We had standard hand-held cameras, another miniature camera concealed in a bag and a vehicle camera, sometimes a video too. The maps carried would usually be of the operation area and about A4 size. They were spotted to match the maps at HQ. We always had a change of clothing, different jackets, another pullover and some sort of hat. The last thing would be the weapons, normally two: a primary, which would be the 9 mm Sig-Sauer pistol and a secondary, which would normally be an HK A1. Only at Camp 3 did we learn how to carry concealed weapons – in civilian clothes, on foot, in public. This may sound easy, you see it in so many movies, but for the first week you think that every man and his dog can see what you have strapped inside your trousers over your right or left buttock. Eventually, you and the weapon become one, as it were. It becomes part of you and you no longer think about it.

Foot surveillance is the hardest to pull off. We would start the basic instruction in the classroom, then move into the camp area. It would be the same with the car, the same rules applied: always have at least one car in support and keep direct visual passes to minimum.

At last we were let loose in the big city. I felt incredibly vulnerable and awkward, not knowing when to move or stay put; I had to be the 'grey man' and yet still communicate on a covert radio – easier these days, you can pretend that you are on your mobile phone! I had to be able to recognise the target without betraying a flicker of recognition. I had to manage to stay in contact with the rest of the team, which often meant keeping an eye on the target while negotiating my way through a crowd, furtively whispering into my jacket. An odd look from a passer-by had no consequences here, but over the water it could be very different.

We were working down south one day; it was an early call and we normally took about an hour and a half to do the journey. Joe and I arrived at our car to find that we had a puncture. Damn. We were going to be late. What the garage did to save time was give us a Triumph 2.5 litre, the vehicle that we used on the driving course – we did not rub our hands with much glee! At that time, it was the car to have – the police used them. We had to get across the country in a hurry.

From the bypass we floored the car right to the start of the M4 motorway. But as we approached the roundabout, a police car flagged us down, all sirens, lights and police authority. We stuffed our maps and guns out of sight and waited for the policeman to approach. I wound down the window. I was going to enjoy this.

'Do you know what speed you were doing, sir?' he asked in best police-speak. (I love the way they say, 'sir'.)

'No, officer.' But I had a feeling I was about to find out.

'We've clocked you at 128 miles an hour. And we couldn't catch you!' He actually smiled. I do like to see people happy in their work.

Out came his notebook.

'Just a moment, officer,' I said. 'Green Dragon.'

'Sorry, sir, what was that?'

'Green Dragon,' I repeated in my most mysterious voice. It was the pre-arranged code-word given us by the training team in case of incidents like this.

He went back to his car and got on the radio. He came back grinning. 'OK, lads, I get the picture. But if you have to speed like that, for fuck's sake be careful!'

We thanked him and he sent us on our way. Two minutes later the sirens and lights were behind us again, only this time the police were doing 130 mph and shot past laughing!

You know that your surveillance skills are working well when you follow a target into a women's underwear shop. If you can overcome the feeling that everyone, especially the target (a woman), is watching you amongst the lingerie, you are on your way to the Det. The shop assistant came up and said, 'And what size is your wife, sir?'

'Oh, the stockings and suspenders are for me, actually,' I said, poker-straight face.

I passed the course. It was a terrific feeling to get to the end. Only 17 out of 128 had made it to go across the water – and Andy and I were among the elite! I was fired up to go now, the prospect of returning to the semi-normality of normal SF soldiering was totally unappealing. The other guys were not SF, just excellent, lateral-thinking soldiers. There were Marines, Paras, Green Jackets and Int. Corp guys to name but a few. Before we departed to the real work we had a week with our families.

Then it was off to work. At London's Victoria station I met 'spy of the year', an MI6 guy wearing (I'm not making this up) a blue three-piece pinstripe suit, pink shirt and tie, topped with a tweed hunting hat. There he stood, reading *The Times* in the corner of the tube station. He ushered us on to the tube for Heathrow to board a C130 at 0200 hours the next morning. The C130 took us to Aldergrove where we were met by the admin staff of our particular Det locations. Later, we were in the bar being entertained by, among others, Gordon, the good friend that encouraged me to go for Selection in the first place. When I was posted to Northern Ireland for the second time, I was confident. Five years down the road from my first trip as a raw Marine, and it couldn't have been more different.

8 Provo Pillow Talk – July 1977

WE SAT IN BATTERED CIVVIE CARS or vans, parked well away from the flat we were watching, hoping none of the locals had sussed us yet. If the IRA knew we were here, we wouldn't just lose the people we were shadowing: we'd probably get ambushed ourselves. Day and night we watched and waited until the Bear Cub, a key 'player' I'd been shadowing, pitched up with a familiar figure by her side.

We rummaged through the photo album we'd been issued with: a rogues' gallery of known 'players'. Some were labelled 'arrest on sight', others were down as 'shoot on sight'. These were the hard core, likely to start firing the second they realised you weren't the pizza delivery guy. A moment's hesitation and your brains would be all over the pavement.

No doubt about it, it was the Big Man himself, at that time commander of a brigade, one of the key figures at the top of the IRA's Army Council.

All hell broke loose on the radio net after we called this in. 'Can anyone go Foxtrot and confirm T1?' asked the OA.

'13A can' was my immediate reply.

'Roger 13A, do you have your camera bag with you?'

'13A affirmative. Wait out.'

'13A to 15B, I will back you. Over.'

'13A roger out.'

I dragged my camera bag from the back seat of the car, checked it was loaded with film and ready to go. I double-checked my radio, hidden in the lining of my jacket. No matter how close the back-up, I'd be on my own for the first vital seconds, if anything went wrong. A comms failure spelt instant death.

Out of the car. Nonchalantly patted down my jacket like I was checking my keys or something. Actually, I wanted to be sure the jacket hadn't ridden up to reveal the 9 mm Browning underneath.

I locked the car (can't be too careful, there may be criminals about) and ambled off towards the heart of town, nicknamed 'Beirut' by the security forces. This was enemy territory. I'd have sweated like hell even in a T-shirt: this was my first 'live' operation in Northern Ireland. I'd just arrived in the Det and I was about to be face-to-face with the most wanted man in Northern Ireland. Without him realising it.

The dialogue in my ear was constant. T1 and Bear Cub were sitting on a park bench on the right side of Beirut. Unfortunately, we did not have long-range directional mikes to pick up their conversation. The only way to find out what was brewing was to walk past, sneak a photo or two, and take it from there.

Whatever T1 did was going to be top news. By 1977 it had dawned on some of the Provos that their hopes of a quick win were an illusion; terrorism alone promised nothing but lengthy jail terms. If your future holds nothing more than a ten-year stretch in Long Kesh you can see why a more sophisticated strategy would appeal. Hence the new emphasis on a political strategy alongside the killings. They'd never get elected like any normal party, and they couldn't shoot their way into power, but a bit of both might do the trick.

So what were the two of them plotting? Nearby, we knew, was an IRA 'quartermaster', i.e. one who hid bomb-making kit, weapons and ammo and arranged for the right hardware to be there at the right time for the next terrorist attack. Would they make contact with the QM, our pretty pair?

I had been part of a team watching the Bear Cub and her associates, 24 hours a day, for 7 months. We looked, we listened. We logged everyone she met and everywhere she visited. Her old white car was allocated the code-name 'Zulu 15' in our radio messages and sighting reports. I knew every dent and scratch and could recognise it anywhere.

Our first OP (observation post) was in a house near her that had stood empty for a few weeks before we moved in. We were concealed in a front bedroom that gave us a great view of the place. Our movement in and out of the place was difficult, but inevitably so as we were right on top of the target. Only one of us could go in at a time creeping into the back of the building. You were in the place for ten days at a stretch.

As the place had been silent for a while, we had regular scares, thinking we'd been bubbled and, at this period, the IRA was as likely to take a pop with an M60 as to quietly reschedule their activities elsewhere. Getting to the house was the worst part, as it was obviously the best time for the IRA to attack us, if they had figured out what was going on just down the road from their quartermaster.

We sat for days on end in a 9 × 5 ft room, equipped with two mattresses, several radio sets and enough camera gear to open a shop in Tottenham Court Road. We had one camera on a tripod fixed on the house opposite and another to photograph up and down the street. We had use of the upstairs facilities, which included a second bedroom and a toilet. We were also free to use the kitchen on the ground floor, though lights were, of course, a problem. Luckily the thick curtains had remained hanging up. But it's lonely, exhausting work; you stare out of the window, wondering what real life is going on around you, staring at the passers-by. Especially the women – most definitely the desirable ones.

The Bear Cub was pretty attractive too, if you could forget what she was responsible for. I felt like a bit of a pervert, logging absolutely every minute of her day. Her fair-skinned colleen face, framed by raven hair, was permanently imprinted on my mind. I could recognise her in the biggest crowd. I memorised her walk, her voice, and by now I probably knew her wardrobe better than she did. After months of intimate observation I almost got jealous when I saw other men talking to her. Oh yes, undercover work can really mess with your head. I was a stalker on government service.

Without any warning, the Bear Cub starting making trips north to the mouth of the Lough. It became a regular thing. What was she up to? Nothing we'd observed so far gave any clue. It made a dramatic change from following her on a shopping trip.

Her white car headed away, and then turned back on itself to head across town.

'All stations. This is 20. Stand by. Zulu 15 about to go mobile.'

'Twelve Alpha. Roger.'

'Fifteen Bravo. Roger.'

'Eleven Alpha and 10 Bravo, Roger.'

If she only knew how many guys were watching her every move.

'This is 20. Zulu 15 with Tango 6 mobile, moving south towards figures 56 Bravo. No, wait. Zulu 15 now going south, cannot confirm whether Tango 6 has company.'

'Fifteen Bravo. Roger I have Zulu 15 moving towards 50 Charlie.'

'Fifteen Bravo, this is 12 Alpha. I'm backing.'

'Fifteen Bravo. Roger.'

'Sixteen Alpha, this is 15 Bravo. Can you take?'

'Sixteen Alpha. Roger. Figures 2 minutes.' Sixteen Alpha was an Army helicopter, trying to fly so as not to make it obvious she was under surveillance.

'Fifteen Bravo. This is 15 Bravo. Roger. I have Zulu 15.'

This was a rush. The news went through the Det. in no time. Every operator wanted to get on the ground. Funny really. Our duty was due to end at 1400 hours. When the Bear Cub moved at 1245, the changeover team was preparing to leave.

She made her way directly to the town, where support for the IRA is notoriously strong. Her home ground but definitely an away game for us. Luckily, it was a beautiful sunny day with cornflower blue sky.

We had to be patient. We didn't know if she was alone in the car or not. Frankly, she'd caught us on the hop by going south instead of north. We had to confirm if she had company.

The car reached the end of its journey. The helicopter couldn't hover without betraying the game, so the ground teams took over.

'All stations. Sixteen Alpha. Two km short of 50 Alpha. Who can?'

'Fourteen Alpha. Roger. I can.'

'Thirteen Bravo backing.'

She stopped outside a papershop, but got back in and headed to the bypass. Was this some sort of anti-surveillance procedure? Or was she going to go on?

We thought she'd go straight through and on towards Belfast, but Zulu 15 made only a short trip instead. She pulled into the central car park (Beirut), stopped and locked up her car – and went window-shopping.

The teams tried to keep one step ahead. Fourteen Alpha was the first there. Park the car, out of the car, lock the car and make sure no pistol barrels or radio wires were showing. He bumbled off, trying for the grey man look, blending in with the scenery.

'Zero Alpha. This is 14 Alpha. Radio check. Over.'

'Fourteen alpha. This is zero Alpha. That's OK. Over.'

'Zero Alpha. This is 14 Alpha. I'm Foxtrot [on foot] towards Zulu 15 and Tango 6.'

'Zero Alpha. Roger. Out.'

Fourteen Alpha was 'Ernie'. A tall, balding guy from the Royal Green Jackets. Excellent surveillance operator, he had a great sense of humour and was well liked in the Det. He was also proud of his green Ford Capri: probably the fastest car in the Det and certainly the most conspicuous.

Was the Bear Cub really just shopping? Forty miles was a fair way for a shopping trip and there was bugger all different here. The mystery deepened when she repeated the trip again and again over the next few weeks. Then it began to come together. Whenever she drove here, whatever else she did, she always checked out some local property. Was she planning to move there? Perhaps she was shagging an estate agent.

She found what she wanted: a flat in a nondescript modern town house, which lay at the end of a cul-de-sac. Crap view for her, but importantly for us, we could get an OP in to cover the place. The team set about fixing up a permanent OP.

We observed the flat without incident until we got a shout from 39 Det (Belfast). The IRA's top men were on the move: none other than Tango 1 and Tango 2. They were driving towards the city limits and surveillance would be handed over to us very soon.

We picked them up on the way into the city, as they turned off, heading south to the apartment the Bear Cub had rented. In they went and we could only twiddle our thumbs and wonder why they'd come all this way.

Our OP was up, so we went in to watch the flat. With an excellent view in, we could keep track of what went on in there and not be seen; if we got any closer, they'd have heard us breathing.

And we didn't have long to wait for a result. We got a call from 39 Det that T2 had been seen leaving with T8, another known member of the IRA's high command. They were headed our way. We picked them up in typical Irish drizzle, following them from the motorway. They drove straight to the flat. No sign of the Big Man.

We settled down at the OP, eyes glued to the binoculars. Finally, we'd discover what they were planning.

They let themselves in, went to the fireplace and switched on the imitation coal fire. They clomped about a bit, made some tea and then ... nothing. They left the room. Had they seen us? We didn't want to have to withdraw. Something was clearly going on, but we were worried we'd been made. Suddenly they reappeared and, turning off the fire, left the flat.

The same thing happened shortly afterwards. And again and again until we realised there wasn't anything wrong with the OP. Had T2 and T8 been in the living room plotting a new act of terrorism, we'd have seen it clearly. But they were in the bedroom. Whenever they went to the flat, they went directly to the bedroom and fucked each other stupid. If there was any terrorism being planned, it was over a post-coital cigarette.

Those two could shag for Ireland, and for seven months they thrashed the mattress. Then came our breakthrough. They came in two cars one time, parked so close we could hear everything clearly, and, when leaving, discussed details of an upcoming 'spectacular'. As far as we could gather, it was planned for nearby. We couldn't believe our luck – we were right above them as they openly talked about it.

'We'll need three people,' T2 said.

T8 kind of grunted. If this was agreement or an indication she'd rather get her kit off again, we couldn't tell.

'We need one as a lookout and driver and two to take out the two policemen or the whole fuckin' police station.'

The targets were drinking coffee next to us: two of the local policemen. They were very cool, giving it 'Oh, ya?' Maybe it wasn't the first time they knew they were being targeted.

We got ready. For months we were on immediate standby every time T2 and T8 came to town. We had no idea of the time frame they were working to, so we had a reception party on the ground for every visit. Yes, including the 'Hereford gunslingers': vans full of camouflaged hitmen just itching to be let loose. If this sounds like bitching, I shall say straight away that in my years with the SBS I've worked with Hereford many times and made some great friends there. However, there are pronounced differences between the two units. Firstly, the SBS recruit from the Royal Marines. So straight away there is a firm and solid foundation for soldiering skills. The SAS does not have this exclusive club to draw on: they recruit from the Army and you can end up with cooks and bottle washers who are extremely fit but who have little or no soldiering foundation. Now this is a slight exaggeration, but it makes the point. Both units have their share of wankers, both officers and enlisted ranks – that will never change. In many ways the difference boiled down to personality. We had it, they did not and we were not going to let go. I spent a lot of time in Hereford for various courses or team tasks. I was used to the steely-eyed 'who the fuck are you?' glare. But I was confident, I had presence, so why the fuck should I let their attitude affect my work. Don't get me wrong, the good people in Hereford and Poole have no need to posture. I was very fortunate to work with some excellent operators from Hereford who taught me many good things about the regiment – unfortunately they're in the minority.

The Det was very different. There we had a tremendous team, the cream of the British armed forces, drawn from all sorts of regiments or corps. Few, incidentally, had any intention of joining an SF unit. We dreaded the regiment's arrival. In a few hours, they could ruin months of our work. Some of them just wanted to provoke a reaction from the 'players'; they deliberately strutted their stuff, hoping for a quick firefight, then fuck off to Stirling Lines and leave us to sort out the mess.

* * *

A few weeks later, T2 and T8 stopped using the apartment. A three-man IRA gun team moved in, as advertised. We watched them cocking and cleaning their weapons. Because, unlike T2 and T8, they weren't bonking all the time, we quickly found out who they were, what they were armed with, and we made preparations accordingly. Now we knew everything. It was time to strike. We could catch them red-handed.

Unfortunately, Hereford was called in. The SAS turned up in their cars armed for a bear-shoot.

We'd asked for permission to make the hit. But each time the answer came back, 'No.' London insisted we hold off. Apparently, Intelligence thought these three particular fish were too small to be worth alerting the Big Man and company to the fact that they'd been compromised.

Eventually, lucky again with the open windows, we heard their leader cancel the operation. 'It's too risky now,' he said, 'there's too many 007s on the ground.' They'd spotted the Hereford presence. Ridiculous and wasteful of everyone's time and effort, but there you are.

The three gunmen left the flat and moved to Beirut – it was Friday and that meant market day. It was a tense time: any wrong move could cost the lives of innocent people in this crowded area. The gunmen moved amongst the crowds waiting for a bus. We knew they were armed. We were in control, but the Det was being badgered by Hereford who still wanted their boys in for a kill. Left to their own devices, we feared they would trigger a shoot-out in the middle of the town, regardless of the danger to civilians. You have to remember that they think they are the only soldiers who can kill. (Most of Det work is intelligence gathering and if we have to draw our guns, we've probably failed in our mission. However, Det operators can and do shoot to kill when their lives or those of the public are in imminent danger. That's the difference.)

The bus arrived, thank God. The gunmen went on to Belfast where 39 Det took over their surveillance. Yes, it was disappointing, there was not a killing, but we'd gleaned a good deal of intelligence about the team and their associates. More importantly, our colleagues in the police station were safe.

It didn't turn out all bad. A few months later, T2 and T8 were arrested. Confronted with the wealth of evidence that they had conspired to murder two policemen, they realised they were facing very long prison terms and no more sex. Well, not the sort they'd enjoy anyway.

They were offered a deal they couldn't refuse. If they told the authorities everything they knew, they would not be charged, but given new identities and set up in new lives out of reach of their former 'comrades'. They sang like canaries. Two of the most intelligent 'players' taken out of circulation and a mass of new intelligence generated. When they'd spilled the lot, they were sent abroad. I know because, by complete chance, I happened to be on the security detail that put them on a plane.

As the wall art says, 'IRA Health Warning: Touting is bad for your health'. They tend to get very medieval with pliers and cigarettes, so T2 and T8 wanted a large ocean or two between them and Ireland.

We staked out the safe house where the loving couple were holed up. We didn't know when they'd be off, just that we were to be on the doorstep, armed and dangerous, by 1800 hours. They left just before midnight. I squinted through the car window, but it was a wet and windy night and I couldn't see them properly. Voices came over the radio and off we went, shadowing their close escort.

We pulled up at a grass strip in the middle of nowhere. The cars went inside and we soon heard the sound of aircraft engines. They were off – and good riddance.

9 Shoot-out in Red Square

I T WAS ONE OF THOSE DAYS when lots of little things had gone wrong – nothing serious, but they added up to a frustrating day. We couldn't know it was about to end in a major firefight. We were maintaining a watchful eye on another IRA woman.

We all knew that she visited Dungannon often; we tended to hang out there most of the time. She rarely went out at night, but this evening she'd driven into the centre of town – 'Red Square' – and parked. She went into a building overlooking Red Square and emerged again about an hour later. This was a new one on us; we wanted to know what she was up to in there. We decided to establish an OP and monitor the goings-on. Anyone going in and out of there would be logged and, if necessary, followed up.

It was a typical fun night in Dungannon: dark, overcast, persistent cold rain beating on the car roof. It's a hard Provisional area, like Lurgan or Armagh town. Not a place to let your guard down. I was tasked with making an entry to this house, to find out if it was suitable for an OP. My oppo had come to the Det from the Parachute Regiment. Joe was about the same height as me, five feet eight inches, but wiry whereas I'm broad. He looked like a meths drinker, his face red and raddled; it looked like it had seen some hard times, which indeed it had. The broken veins in his nose, the beetroot cheeks, receding ginger hair – and the fact that he smoke and drank in heroic quantities – made him pass very well for one of the locals. He was a very professional operator, and I got on with him extremely well.

Going in, we had a number of car teams to provide us with back-up. Some were parked at a distance on instant alert, others drove around the area, keeping an eye out for the opposition. Our car, with our driver, 'Noddy', was parked right outside Red Square.

I tensed the moment we broke into the house. It felt wrong. Nothing I could put a name to, but I sensed an atmosphere here. It had been abandoned for some time and it was an eerie place to creep about at dark o'clock. Shadows. Half-shapes. Loud creaks when neither of us had moved. From a security point of view it was fine; we were just glad we wouldn't be the ones holing up here.

Our driver had joined the Det from the Green Jackets. Noddy was a tall lad with a long, loping stride and tonight he was sitting at the wheel of one of his favourite cars. His gold Vauxhall Cavalier GL was comfy to sit in for long periods, as we often had to, and it went like shit off a shovel. But I hated it. The Cavalier is too open: acres of glass that enable anyone outside the vehicle to see exactly who's in it and what they're doing. I also thought the RUC and the Det used too many Cavaliers, they were danger-ously close to our trademark. That's why, on the principle of 'if you can't beat 'em, join 'em,' I'd captured myself a Datsun 120Y. Mine was a vile shade of brown. With its narrow windows, sharply sloped roof and big fat head-rests, it was extremely difficult for an observer to see who or what was inside. And doubly so for a short-arse like me.

We crept out of the building and made for the end of the garden. Ahead was a six-foot wall that backed on to the square. I had a bad feeling in my gut before we started to climb over. We'd been in there too long. As we dropped over and back into the square, the first shots rang out, sharp reports, loud despite the rain.

I hit the ground. I couldn't see a thing through the drizzle. The only thing that was clear was that we weren't the targets. They were shooting up the Cavalier. With Noddy still inside . . .

Then I saw them. Two rough-looking characters in greatcoats, long greasy hair slicked back by the rain (this was the standard IRA 'look' in the late 70s). They had what looked like a 9 mm and .38 mm revolver or it could have been M16s in their hands, which made them fair game. I stood up, drew my Browning and squeezed the trigger.

'Contact front!' I bellowed.

Joe fired moments later.

The Browning is an accurate handgun, but we were shooting at extreme range for a pistol, a good 150 feet and it was dark and raining. We 'pepper-potted' forwards (one man moving while the other man covers). God, I wanted these bastards! Time they had a dose of their own medicine.

I put out a contact report over the radio as we pursued them. I had an engine over-rev, then a screech of tyres as Luke appeared from down the hill, slewed his car across the road and flew out of the door, gun in hand. Lights went on in the neighbouring houses; doors opened and some people came out to gawp – or worse.

The gunmen reached the far wall. They were fast and fit and would be over it and gone in an instant. I took a two-handed grip, side-on, my left hand locking into my right to provide as steadier platform as possible. Over the roof of the Datsun I squeezed off a series of aimed shots. But I couldn't concentrate on the shot. I had a wounded mate, probably dying, a few yards away. And in his car were his weapons and our operational code books: gold-dust for the IRA.

I put five or six shots down at the right-hand gunman, then emptied the rest of the magazine at the other. I saw the second man flinch.

'Changing mag!' I yelled, but before I could get another one in the pistol, the gunmen had made it. They were up and over the wall and we couldn't go after them. We had to see to Noddy. Hearts pounding, we skirmished over to the Cavalier.

It was a sorry state. The windscreen was shot to pieces. So was Noddy. He was slumped unconscious over the passenger seat. But he wasn't dead. He'd been incredibly quick-thinking – and incredibly lucky. As they'd opened fire at him through the glass, he'd thrown himself over and to the left, away from the main cone of fire. But the first five or six shots had caught him in the right side of his chest blasting it apart and he had been hit in the mouth. There was blood, glass and bits of him everywhere.

Luke was medically qualified and was on the scene when we arrived. He took one look at the massive chest wounds, jumped into the shattered Cavalier, and drove off at top speed. If Noddy wasn't on an operating table in the next few minutes, he'd be delivered to the morgue. We got in Luke's car and followed him to the hospital.

Luke got Noddy up to casualty, shoved him in a wheelchair, and pushed him straight through to the operating theatre, shouting for help all the way at the top of his voice, his power of persuasion with his 9 mm was so impressive that the head nurse help him save Noddy's life; her husband had been killed by the IRA two years earlier – she did understand. His action saved Noddy's life. The whole of his right-hand side had to be rebuilt, but he lived. Years later he still wakes up in the middle of the night to see the glass explode in his face, and feel the bullets slam into him.

On a happier note, later that night the RUC nicked a long-haired man in a greatcoat who pitched up to Musgrave Park Hospital with a 9 mm bullet in his leg.

Coming back from the Det proved almost as hard as training to join it. I was posted back to the SBS, ready for maritime counter-terrorism (MCT) work. As a Det operator, you spend at least 90 per cent of your time on your own. You may come together with others to do a recce, or to put in an OP, but day after day, night after night, you're in that car, watching. And you're alone. Solitude and solo pressure does certain things to you. You look out for yourself to the exclusion of everyone else. In fact you can become so driven into yourself that other people, even loved ones, effectively cease to exist.

It's the little things that trip you up, betray what you've become. I'd go home on leave, finally, and make a cup of tea. But just for me. Not for Fiona. I'd always been polite, but now I found myself going through the door in front of her, not even noticing she was there. She said nothing at first, just giving me time to sort myself out. But by God I tested her patience! Going out was the worst. Once I actually got into the vehicle and drove off, forgetting she was there. Then I realised I'd left her on the pavement and had to go around to pick her up. She rightly went ballistic and told me to extract my head from my navel. Fair comment. Too long on the Det can really screw you up inside. It took some time to get back to a normal life.

10 Spying on a Russian's Bottom

Aᴜᴛᴜᴍɴ 1978. There's a strange tension at Poole, an air of mystery and intrigue that meant someone's let something slip from the operations room. The rumour mill was in full flow, speculation getting wilder by the day, but none of the fanciful suggestions were nearly as far-fetched as the operation we eventually undertook.

Various guys were selected for a team, but those in the know remained tight-lipped. Even the initial briefing, when it came, made no reference to the real operation. All that came from Den, the operations officer, was a Warning Order that left out both the area and the perceived task. The odd thing about this particular operation was no existing team was selected; they assembled a mixture of talent from the three different squadrons.

Looking around at the final group, the team was made of talented, fit and totally professional individuals. I had never seen such painstaking selection of people to conduct a job, and I've never seen it again since.

Paul and I were the youngest of the nine-man team. We had always got on well since surviving the same Selection cadre. We've often worked together in our subsequent careers. Barrel-chested, extremely fit and a little taller than me (not difficult), Paul's dry Brummie humour and robust good looks always got him noticed. With only five years' service in the SBS he was involved in the planning and preparation of this operation. His dedication to the Service was second to none. (Tragically, at the peak of his career, he was killed conducting experimental demolitions training. A terrible loss to us all – the most respected within the Service and one who will always be remembered.)

The team was headed up by Glen, who emerged from his briefing with the CO with an air of conspiratorial glee. Down the corridor he came, past the wall covered with memorabilia from past operations and photos of the Old and Bold, and into the operations room.

'OK lads, you know the score,' he said, piercing eyes looking straight at me and Paul. Yup. It was going to be one of those, 'I can tell you but then I'll have to feed you through the paper shredder' conversations.

'No words to anyone, I repeat, not one word and that includes family and friends. Emergency phone calls only. You will now get all your diving equipment and be ready to move at 1800 hours tonight. You will be required to be self-supporting for about six weeks with reference to money and civilian clothes.' Civvies? Where the hell were they sending us?

'Oh, and do not forget your passports.' This was getting better by the minute.

'Are there any questions?' Nothing printable or relevant occurred to me.

'No, good, see you all tonight at 1800 hours.'

Is that it?

Paul and I looked at each other in amazement. 'What the fuck is this all about, Paul?'

'Nothing, mate. Let it run and see,' was his typical answer. We were supposed to be ready for anything, but you can take this cool professional thing too far sometimes. We walked out of the operations room, down the long corridor outside and towards two SBS lines. Paul and me walked into a barrage of questions from the team. They did not believe that we still didn't know what the mission was, and assumed we were just trying to be clever.

By 1730 hours piles of kit were assembled in the compound as the team came together. Paul and I lugged our bags out ready for the transport to take us away. The remainder of the team – Glen, Den, Mike (known as 'Popeye'), Ron, Bert, Simon and Tim – turned up in their cars at various intervals, driven by their wives. Everybody had their own way of doing things in this family situation. Some of the wives were visibly upset and just wanted to drop off their man and go home; others lingered, unable to let go so suddenly, much to the embarrassment of their husbands. The team was together, but still no updated brief from Den. Then the transport arrived, kit bags were slung into the wagons, some hurried goodbye kisses and we were rattling our way to RAF Lyneham.

We flew Crab Air to Gibraltar, the RAF Hercules living up to its legendary reputation for mid-air comfort. Gibraltar has always been a favourite hunting ground for the SBS. Up until the mid-1970s there was always an SBS team there on standby. It's the gateway to the Mediterranean and Middle East, and General Franco was always rattling his sabre, making dire threats to seize the place. But he died and the Labour government withdrew the team, just as they pulled the lads out of Bahrain in the sixties. Come on, guys, where would you rather be stationed, the Med, the Middle East or Dorset?

The Rock of Gibraltar has to be seen to be believed. Connected to Spain by a short flat strip of land, it slopes up almost vertically like a gigantic wall. There's no way of storming it from the front. Tucked around the corner is the harbour and a fleet based there can dominate the straits, controlling traffic between the Mediterranean and Atlantic. Captured by the Royal Marines three hundred years ago, it's been a key British naval base ever since.

The SBS has conducted many operations from Gibraltar, using it as a mounting base for operations. The Service still used Gibraltar and HMS *Rooke* as an advanced diving training ground and continued to train there often. In all cases we have an excellent rapport with the Navy and HMS *Rooke*. So it would be no surprise for an SBS team to pitch up there, but Den stressed to everyone that we must maintain the usual low profile. Usual? We laughed at that one, but we got the point.

For four hours the Hercules droned over Biscay and round the corner to the airstrip that straddles the neck of land north of the Rock. (The Spanish, still sniffy about surrendering the place, wouldn't let us overfly their territory.) Den sat there with his nose in a red folder marked 'Top Secret'. They really do have them.

We'd had enough of mists and mellow fruitfulness of the British winter: in Gib the sun beat down and it was T-shirt weather again. We moved on to HMS *Rooke*, established ourselves in familiar surroundings and were told by Glen that we would not be required for the next two days. Den, Ron and Glen disappeared with the head shed for endless briefings, while we soaked up rays and generally chilled out. Rank has its privileges but I've got the tan. We saw them at intervals, looking really intense, but unable to tell us anything yet.

We hung out at the Buccaneer, a typical English pub just over the road from HMS *Rooke*. We moved our kit to the diving shed on Coaling Island from where we would initially work. However, as the operation took shape we were moved to the South Mole which became our operations room and planning area.

HMS *Rooke* is in the centre of the town, directly on the harbour front, all very picturesque. The Navy police sighed as we entered the establishment. They knew once again their patience was about to be tested. Although standard procedure in SBS life, some of our routines were frowned upon. A simple thing like physical training in the morning, running through the gates at 0615 hours, was a no-no to the naval establishment. But we didn't have a problem, because they couldn't catch us! Once through the gates the road forks; to the left is the officers' mess, in the centre is the senior rates' mess and to the right the other ranks' accommodation. The buildings were white apartment blocks four storeys high, looking across the harbour towards the Detached Mole and harbour entrance of the South Mole.

HMS *Rooke* brings back so many memories. We have a reputation of working hard and playing hard. Our work has been second to none, the social scene within the messes just as good. Each mess from the officers' mess to the senior rates' mess has experienced our good will. From the men in black abseiling from the roof of the officers' mess to deliver a box of Milk Tray during a cocktail party, to walking through the plate glass conservatory doors in the senior rates' mess (actually, we just didn't see the bloody things). All now part of the legend of HMS *Rooke*.

What we didn't realise was that if we cocked up this mission, the HMS *Rooke* would be on front pages everywhere. Gibraltar wasn't just the base, it was the scene of the operation itself!

Den's briefing began with a history of this type of mission. Apparently, in the early days of the Cold War, Soviet warships still visited British ports. Navy divers, in league with British intelligence, would have a discreet snoop around them while they were in harbour. A close inspection would reveal all sorts of information about their onboard sensors, weapons and propulsion systems, likely top speed and so on. Unfortunately, it all went horribly wrong when Commander Crabb, an experienced World War II frogman, vanished in Portsmouth harbour while having a secret peek at a Russian cruiser. Weeks later, his headless body washed ashore, the media got wind of the story and the Soviets went ballistic. No one ever knew what happened to him. Had the Russians sent their own divers down to guard the ship? Had he just met with a dreadful accident? Where was his head?

Our operation turned out to be a repeat of Crabb's. In the late 1950s these operations had been very amateurish;

the hope was, with our professionalism, we'd be able to get a result and no one would be any the wiser. It turned out that with détente in full swing, Anglo-Russian relations were warmer now, and one of their new Kirov class battle cruisers was going to make a port visit – to Gibraltar. We were ordered to conduct a hull recce of the cruiser as she entered the harbour and while she was berthed alongside. Den assured us that we would know in advance exactly where the Russian warship would moor, and there is only one way into Gibraltar harbour. No problem there. Then he passed around pictures of the Kirov. Taken from all angles, left, right, rear, front and vertical; the Nimrods and Tornados had done a good job, but there was, obviously, nothing from underneath. That was down to us.

The brief continued with the need for high security on what we were doing. But it ended on a complimentary note. 'Gentlemen,' Den concluded, 'you've each been selected for your own personal skills. You are considered the elite of the elite – so please act like it!'

We gathered the pictures and paperwork and gave them to Glen. Den left for a meeting with the Admiralty high-ups. This was a high-risk operation, monitored by the Prime Minister's office. If any of us were to be caught or killed the political fallout would have been extremely embarrassing. We were reminded of the fate of Gary Powers, the American U-2 spyplane pilot shot down on an illegal overflight of the Soviet Union during the 1960s. He'd survived the crash but served ten years in a Russian jail until they cut a deal to let him go.

The Russian cruiser was due in port in four weeks' time: D-Day was the first week of May. The first decision was to relocate ourselves and our gear from Coaling Island to a new building on the South Mole. This was an isolated red brick warehouse near to the entrance to Gibraltar harbour. The building (no. 104) was one of several Navy store-houses on the mole, some of which were now disused. It was vast, our nine-man team lost in cavernous space, our voices echoing around the place. The roof was 50 feet above the floor. The concrete floor was cold. The good thing about it was that it offered a relaxing break from the hot sun outside. There were several separate areas within the building that would make us self-sufficient. We had a galley and eating area with all the facilities that we required. Behind the galley was our sleeping area and within the centre of the building was the Operation Room.

The main entrance looked out across Gibraltar harbour with steps leading down to the water. To the rear of the building there was an entrance that led to the other side of the harbour wall and the Mediterranean Sea. It was decided that this is where we would live, eat, play, plan and work. We would miss the comforts of HMS *Rooke*.

Our preparation began that day. Each man checked and tested his kit, then double-checked. Meanwhile, the planners and developers sat around the table and discussed the way ahead for the following four weeks. Only then did it really sink in what a delicate operation this was. There was a nervous joviality amongst the team, the classic sign of confidence combined with extreme apprehension. Even with my lack of experience I realised that we were all confident we could do it. And there's nothing wrong with being nervous.

We were to pit our wits against the world's most secretive navy, whose latest and most sinister warship had already attracted feverish speculation from the defence press. In classic spook manner, we sanitised our kit, removing any label or item that could be traced back to the UK. I had my doubts about this: I mean, if they caught us, who the hell would they take us for? The CIA? Jacques Cousteau's mini-sub crew?

On completion of our equipment checks we made out our personal kit lists. This was a standard operating procedure that we conducted for any task. We make a complete list of what we intend to wear during the operation, right down to the colour and design of your underpants.

(Best get rid of that St Michael label then.) This list would change frequently over the next few weeks, as the training progressed.

Sometimes you could enter the water and feel good. Other days you would be unbalanced and your trim would be wrong for what seemed no reason, until you went back over the changes that you had made to your kit since the last dive. It's amazing how much difference a different undersuit or additional bit of kit could make. I wrote down every combination of equipment, noting what effect it had. The aim was to be as smooth and comfortable underwater as we could be.

We swam on normal compass swims; whether we swam with compasses or cameras, it would be necessary to know exactly what weight was required to enable a good trim at 20, 23 or 26 feet.

Glen allocated the proposed team pairs for the operation. These teams were up for discussion, but since everyone got on with each other, we were happy to go with Glen's choices. If there was an injury to a team member, one of the reserve team would take his place. Paul and I were chosen as a pair, which suited us as we had done many of these types of recces in training and understood each other on the surface and under the water. The team breakdown was as follows: Den H – operations officer, Glen – team leader, Ron W – team sergeant major and the swim pairs would be Bert/Popeye, Simon/Tim and Paul/ myself. There had to be a reserve team and that team were named as Bert/Popeye; this was accepted by all concerned. Glen's main worry was security. We were working from a familiar area where the whole of the island knew that it was now the new home of the SBS. So whatever we used during a day's training had to be removed and reinstalled the following day.

Every day began with physical training. This varied from day to day, but greater emphasis was placed on running as that would take us in sight of the proposed mooring site of the cruiser. We ran from HMS *Rooke* to the South Mole, swam from the South Mole to the Detached Mole (where the cruiser was to be berthed) and then to the North Mole and then back into HMS *Rooke*. This added up to 8 miles of running and about 500 yards to swim in open water. After that, time for a humungous, greasy Navy breakfast, eggs with everything. Diving training followed.

We undertook two dives a day and a third at night. The temptation was to dive around the intended berth of the Kirov, but that would have been a giveaway. We had to assume that the KGB was already in town, with guys watching the harbour front with high-powered binos. We made the occasional dive around the Detached Mole, but our visits to the area were mainly reserved for night dives, when the guys in fur hats wouldn't be able to see anything. We swam unmarked, no safety floats, only our safety boat knowing where we were. We got to know every fish and sea plant in the area and felt sure we could swim there blindfolded – well, maybe. Gibraltar harbour was at most times an enjoyable place to dive. The water was clear and the natural harbour depth was some 60 feet.

Paul and I entered the water from the steps of the new building. We surfaced and gave a thumbs-up to Bert in the safety boat. We checked each other for any high/low pressure leaks from the SCBA, ensured that everything was secure and surfaced to inform the surface party we were OK. Today's dive was a standard compass orientation dive of the immediate area. We had a compass course to negotiate and five compass points to visit. We looked at each other, gave the thumbs-up, and dived. It was as if we had been in the water together all year. We settled down to the prescribed depth. Paul took the first compass leg, so I settled just above his left shoulder, looking down on the compass board which contained the compass, watch and depth gauge. It was crucial that we maintained a good depth, which on this dive was about 24 feet. Our trim was good, we weren't bumping into each other and our leg kicks were helping to maintain a good straight line. As the no. 2 on the swim team, it was my responsibility to look after the buddy line. This is about 13 feet long, attached to Paul's left upper arm and my upper right arm – this was standard throughout the training which enabled us to have standard swim positions every dive. The buddy line was also used as a measuring instrument, and we'd use it to take rough measurements of anything we encountered – like a Russian cruiser's hull form. I held the buddy line coiled in my right hand and the float line, which led to the surface, in my left hand.

The first leg was 550 yards and was to buoy 56, located in the middle of Gibraltar harbour. Every 100 leg-kicks through the water would be about 325 feet and took about 1 minute and 30 seconds. So the 550 yards would take about 7.5 minutes. Within that distance we would undertake at least one 'observe'. This is a technique used to check your course is correct. In this case, Paul indicated that he was about to undertake a quick observe by pointing to his eyes. We would maintain our swimming but gradually close the surface; I would maintain a depth of about 16 feet as Paul continued to the surface. His face mask would be the only thing that broke the surface and, maintaining his compass bearing as he did so, he would observe the target: in this case buoy 56. As long as we were on target, he would descend and carry on swimming. At this stage I would be swimming below him and easing him down to my depth. A quick tug on the buddy line would indicate that all was well and we would carry on. The only indication on the surface would have been the slowing of the bobbing float. The surface safety team might catch a glimpse of mask. It was a technique that needed practice, but once we'd rehearsed a few times it went like clockwork.

The rattle of the chain told us we were getting close to the first compass check point. There it was, the chain dangling down from the buoy to the seabed. We clung to the chain and swapped the compass board over. Paul attached it to my low-pressure tube on my SCBA. I broke the surface, using the buoy as cover, and set the compass for the next leg to the Coaling Island jetty. Off we went again, this time with Paul on my left side.

We arrived back at the steps hitting the sea wall to the left of the steps, shit, between ten and sixteen feet off target. We surfaced and acknowledged the surface safety party and they instructed us to leave the water.

'Good dive, lads' was the shout as we removed our hoods and masks. We knew it was good, although there were a few little things to sort out. We were out of the water after 90 minutes with a spring in our confident, cocky steps. It will be perfect by D-Day, we said.

The training went well. We dived at night twice a week. There was always a ship available on which to conduct training. Royal Navy warships needed to exercise the state of readiness they would be in if there was a threat from underwater swimmers, in order that they passed the necessary procedural tests set by the Admiralty. At this state of readiness the ship conducts a number of measures to counter the threat of enemy frogmen. These include turning on underwater lighting, scanning with sonar, setting the underwater suction vents on full power, and sending down clearance divers to check the hull. Small explosive charges might be dropped over the side and the ship's propellers could start turning at any moment.

The odds are stacked in favour of the ship and it can be a very dangerous place to dive. But then, no navy wants to lose £50 million worth of warship to a diver with a £50 limpet mine! The only thing in the divers' favour is that we get to choose the timing of the attack.

Our first training ship attack was against HMS *Plymouth*, a frigate that had just arrived in port after six weeks at sea. They were not allowed alongside until they had conducted their underwater-threat exercise. The ship was anchored off our old friend buoy 56 in the middle of the harbour.

Briefed at 1400 hours, we entered the water on the seaward side of the South Mole. We had to swim north along the South Mole, through the harbour entrance, then on a compass bearing to buoy 56 and HMS *Plymouth*. If our luck was in, our return route would be the same; however, we would finish by swimming to a boat some 500 yards off the entrance of the harbour. As usual, the team talked in detail about every aspect of the route in, our actions under the ship and finally our extraction route.

To our horror, Paul and I were told we'd be captured – however well the rest of the mission went. The ship's company had to test their interrogation techniques too: just in case their defensive measures fished out one of our Russian opposite numbers during a real war. We got the short straw, being the youngest. The other guys would head back for a shower, while we got the 'for you the war is over' routine. And you never knew how far people would go, even if it was 'only an exercise'.

Preparation and briefing complete, we entered the water from the rear of our new building. A bit different to walking down the steps and into the harbour itself. We had to negotiate the rocky side of the South Mole and the Bay of Gibraltar. There was a total of three pairs conducting this 'ship attack' and 'hull recce'. We all had a specific job to do. Paul and I would do the hull recce while Bert and Popeye would conduct a ship attack and place their 'limpets' on the stern of the ship. Divers C and D would do exactly the same to the bows.

Into the water and a thorough check for leaks and anything else that may need fixing. A polite thumbs-up to the surface safety and we were away. That would be the last that they would see of us until we reappeared after the extraction swim. All they could do was watch and react to any emergency. This is what we called an unmarked swim at night; all the boats could do was guesstimate where we were and manoeuvre the safety boats to cover our route. The safety boat carried a diving supervisor (responsible for the divers and diving procedures), a coxswain (responsible for the boat, engines and boat stand-by equipment), a stand-by diver (his responsibility was the divers in the water during emergency procedures – supervised by the diver's attendant and controlled by the diving supervisor) and a diver's attendant (to help the stand-by diver during emergency). During diving operations/exercises the boat would show either a 'Flag Alpha' during the day or 'Diving Lights' at night. In this case we had permission to show no diving lights unless there was an emergency. It made the exercise much more realistic.

We had to do everything by feel. You couldn't see anything under water apart from occasional phosphorescent plumes streaming back from the compass board as we swam along the south edge of the South Mole. We realised we were nearing the harbour entrance, the tide propelling us onwards. Twenty-three feet down, we sensed the change in the weather topside, the swell much stronger than this morning. The met report had indicated an approaching low pressure front, but not this early.

I felt Paul's hand on my arm: we'd reached the entrance and were ready for the next leg. The swell moderated and it was easier to maintain depth. Our plan was to take a good 'observe' of the target on the other side of the South Mole. Paul surfaced again using the wall and weeds as cover. I was sixteen feet below, waiting for the tug on the buddy line. The line leaped in my hand. I pulled him down and we continued the swim, the final leg. Two thousand feet to go.

We'd got half way when they turned on the lights. *Plymouth*'s underwater lighting cast an eerie glow even at this distance. Had we been spotted already? The area around the hull would be lit up like a Christmas tree. Her boats were out too, circling round to protect the mothership. Noise travels far and fast in water, and it felt like they were right above our heads.

There was only one thing to do: ignore the safety regulations. We'd discussed this before and agreed amongst ourselves that if the lights were already on, we'd just have to go deep. Our maximum safe depth (MSD) limitation on oxygen was 26–32 feet, but we knew the French navy divers routinely went to 65 feet. It was illegal under our regulations for the very good reason that you could very well go down with oxygen poisoning. However, we felt we'd be OK. We weren't going to be that deep for long.

You'd think it couldn't get any blacker, but as we dived steadily deeper, it became utterly disorientating. With a soft thud, I swam straight into the mud at the bottom of the harbour. My diving suit squeezed me all over as the pressure soared to three times what we'd been supposed to be under. I heard a hiss as Paul opened his bypass valve to let more oxygen into his counter-lung.

I could see a glimmering light up ahead, seemingly miles in the distance. It was the ship's illuminated hull, in fact no more than 600 feet away. Dull pops sounded as if someone was dropping explosive charges too. I wondered if their sonar operators heard us open the bypasses.

We swam slowly towards the light until we were directly beneath the *Plymouth*. As we neared the illuminated zone, we pressed ever deeper. Screw the regulations, we were going to get a result tonight!

Now we could hear the noise of a live ship, machinery running, metal on metal, but how close were we? The sounds got louder. I put out my hands, expecting at any moment to rest my palms against the warship's solid sides. At last! My fingertips brushed the hull.

Paul had a narrow escape. He swam into an opening in the hull, an induction pipe that is supposed to be covered by a mesh cage to stop the system ingesting debris and, er, divers. If it had been switched on, he would have been sucked into the machine. Thank God it was off.

We didn't have time to dwell on his brush with death. We got straight on with the hull recce, sliding along the *Plymouth* from bows to stern.

Suddenly, something grabbed my right leg.

What the fuck? I wrenched it free, gasping for air, blood pounding in my ears. It was Bert and Popeye: they had hidden themselves on the propeller shaft, waiting for us to swim past. Just couldn't resist it, the loveable comedians. Of course I can take a joke. I swore to get the bastards back, first chance I had.

We completed the hull recce and measured every inch of the ship using our buddy line, which was exactly 13 feet long. We were looking for the unusual signs that might indicate a different type of sonar or variable prop shafts that could give the ship greater manoeuvrability or higher speed. Although ships like the *Plymouth* were a tad small for this type of operation, it meant that we would go through the procedures that we do on a live target.

We completed the recce and swam out 200 yards before we reluctantly surfaced to give ourselves up to the crew of the *Plymouth*. Ship's boats circled us like bees around a honey pot, then we were dragged from the black water. The Navy lads had obviously experienced this many times and although we had a great rapport with the Navy, this was their one chance to get even with the elite commandos, and fuck! Did they get even! We were ordered to lay face down on the bottom of the boat and our hands were dragged behind our backs. Our diving sets had not been removed so it was very uncomfortable until our sergeant major came to the rescue – not for us, but for his equipment that was on his slop chit!

We were dragged up the moving stairway on the port amidships side of *Plymouth*. Black hoods were shoved over our heads and, although I could see shadows, it was a case of being led or falling down, which happened several times. We were spreadeagled on the cable deck port side, stretched to our limits. I could see from the bottom of my black mask that there was obviously some sort of social function in progress: the officers were in their cocktail suits looking rather smart – we were their dessert and after-dinner entertainment!

In Special Forces, we were taught that if you are going to escape you must go early before they (the enemy) establish such a routine that escape is near impossible. I noticed through the bottom of my headgear that the officers had encircled us as we were being stripped. Not too far behind me there were two naval officer types who tended to get excited as the wet suits were ripped away: they were standing next to the wire rail that led to the water some 35 feet below. I kept on saying to myself, I must go now before it all gets too late. I kept on pushing myself trying to find Paul's hand to indicate my full intentions. No luck: he was some 15 feet to my right. At that point I turned and ripped my headdress off. With my wetsuit bottoms still on, I charged towards the only way out.

I pounded through the immediate circle of officers, heading for the railing. Before they could react, I seized both officers in their cocktail suits and dragged them over the side with me. Even at the time, it felt like slow motion: the look of total astonishment on their faces as I grabbed them and heaved; then terror, as they realised I had the upper body strength (and momentum) to topple them overboard. We seemed to hang in the air for a particularly long time before we hit the sea.

Only later did I learn that these two officers thought I'd do something really nasty to them once I had them in the water. In fact, nothing was further from my mind. My plan was to swim to the keel of the boat then forward to the bow. It worked like a dream. I made it to the bow with all the ship's attention focused on the rescue of the two unfortunate officers. There was a lot of shouting from the officers in the water – their bow ties were shrinking around their necks and they could not do anything about it! (I don't recall James Bond having this problem.)

Unfortunately, no sooner had I achieved my escape when the sergeant major, knowing exactly what I would do, appeared at the bows and shouted: 'Don, get in the boat.'

I'd hoped to avoid the R to I (resistance to interrogation) phase with my brilliant escape.

'Don, you've got to go back so they can exercise their routine.'

Bollocks. I was gobsmacked. All that for nothing. I trudged slowly up the gangway to where Paul and myself were to complete the cycle. I reached the top of the gangway to see the officers, now in their wet cocktail suits, looking seriously pissed off. A wry smile crossed my face as I made eye contact. Oh dear, they couldn't take a joke. I was kicked to the ground. My wetsuit was now completely removed and my little Willy Wonker was like a walnut whip. However, this seemed to excite my fan club of the two young officers. I was to see these two quite frequently over the next twelve hours.

My mind drifted back to my instructions for situations such as this. As a Special Forces soldier, operating behind hostile lines, you may find yourself one day in a scenario that is nightmarish to say the least . . . That is, a scenario in which you are taken captive and held by your enemy. If this does happen, it was important that you are aware of the agreement which has been put together between nations to protect the fundamental rights of all prisoners of war: the famous Geneva Convention. We were taught this until it was second nature. Obviously, I hoped it would not have to be put into practice except on the odd exercise or escape and evasion (E&E) scenario.

The Geneva Convention is an international treaty that covers all aspects of the treatment of prisoners of war (POWs). It was designed to protect the fundamental rights of all prisoners of war, regardless of race, nationality, political or religious belief. Most western countries honour this agreement; however, third world nations do have a habit of not abiding by the rules.

The Convention gives the senior ranking prisoner being held a legal basis under international law for his demands for humanitarian treatment and the necessities for a decent and honourable survival for all prisoners. Every soldier should have a working knowledge of the accords as it could save his life or, at the very least, make his existence more bearable until he can escape or is legitimately released. The most important subject of the Convention is captivity and the major elements are covered in articles 17–118. There are various areas covered which include: the fact that POWs are required to give only their name, rank, service number and date of birth. The use of physical coercion to obtain information from POWs is prohibited, but all nations will employ some level of duress to extract more information. The more important the information, the more tempting it is to give the prisoner a kicking. And since Special Forces guys are liable to be up to something highly secretive and have access to unusual kit, our fate in enemy hands is unlikely to be pleasant.

It was stressed to us that it was important for you, as a Special Forces soldier, to understand your rights to fair treatment as a soldier in the event you are ever taken prisoner. Remember, it was going through my head, I am only required to give: name, rank, service number and date of birth.

I was hauled up on the ship's crane, stark naked, with just hessian cloth face cover over my head. As they squirted me with water from a high pressure hose, up my butt and between my legs when they aimed straight, I wondered, does this come under the good old Geneva convention? The water was freezing cold, the icy blast so powerful that I spun around like a cork. Knowing this information could possibly save your life or at the very least, make life in captivity bearable until you can escape or are legitimately released – what a joke. I was just hanging there, the crowd silent; I could just see out under the mask and spotted my young officer friends who were stood there in a daze with their erections as proud as anything for all to see – not a wee bit embarrassed about the whole thing.

Above The author, Jungle training, Brunei. This was the first 'joint training' selection when the SAS/SBS units came together

Left 'Hot Extraction': Jungle evacuation by helicopter

Above and below Oman: preparing for the final push through Raykut and Delkut, and on to the Yemen border, Huey UH1Bs and Strikemasters in support

Above and below The Hawker Hunter offers air support to the land assault

Above The LZ – Landing Zone – for Hueys in Oman. The Yemen border is within sight

Below My turn at the wheel – the inaugural flight of Camsell Airways!

Right Air-to-ship assault . . .

Below . . . and the back-up troops fast-roping on to the target

Above As we take our positions in the 'lurking area', the team are in the final stages before submerging

Below Dive Dive Dive – the beginning of the run to the MCT target

Above Old technology: the older yet well-used Long Range Insertion Craft (LRIC) as the team makes a final approach

Below The arduous climb to board a target

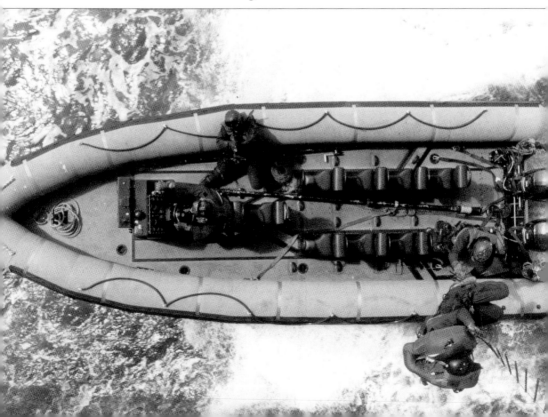

The 21st century – the new LRIC

Suddenly, I fell like a stone. I braced myself to hit the ground, but the bastard crane operator stopped it just at the last moment. My shoulders and back were wrenched hard and I shot back up in the air. When they got bored with that, they swung me out over the side and dunked me in the sea. I dreaded being keelhauled – pulled from one side of the ship to the other. You see it in the movies and it was something that I was not particularly fond of. Luckily for us, that had not crossed their minds. They just wanted to drown me. They left me under water for ages; I passed out a couple of times before coming to, coughing up water while still dangling from the crane. Then they'd dunk me again.

I have no idea how long Paul and I were dunked in the harbour, but eventually they brought us on to the deck. We were shoved inside 50 gal drums, full up with ice and water. And left to contemplate the errors of our ways.

It's all to soften you up. You're freezing, hungry and very tired. Then you're dragged into a warm office and the cuffs are removed. They handcuff your hands in front and pull off the hood. You stand there naked, in front of a table with a man dressed in naval uniform, pen in hand, paper on desk. It's your invitation to tell all so you don't have to go back to the drum of icy water again.

'Name,' he said, without even looking up.

I stuttered, trying to calm down and get my mind in order. The guy looked up and politely asked again.

'Name.' He was obviously the 'nice' guy; he didn't shout throughout the 30 or 40 minutes I was with him. I was adamant that all he was getting was my name, rank, number and date of birth. Even in the warm room, I was shivering.

'Would you like a cup of tea?' he asked.

'Y, Y, Y, Yes.' I wasn't faking, I couldn't talk properly until I thawed out.

The bell rang and, a moment later, the steward moved in with the tea. I was gagging for a drink. On the R to I course, we're told, 'take everything that is given to you'. I intended to do just that. The tea looked so good as the steward placed the cup in my hand. It was warm and inviting. I could smell the sugar. But just as I raised the cup to my lips, the door crashed open and knocked the cup from my hand. Paul was thrown on the deck. I was grabbed and dragged back to my ice bucket.

Surely they must stop this shortly, I prayed; I didn't even get a mouthful of tea. Fuck. We went back to that office three more times and the result was the same: name, rank, number and date of birth is all they got. You could feel the annoyance in their voices as the night went on. I felt good, but I was still concerned at what seemed an eternity as the exercise continued. Then it dawned on me: if the Kirov operation went wrong, I could be doing this for real next week, with real live KGB men ready with the electrodes. I must be out of my tree.

The water was so cold it hurt. It reminded me of my worst time in cold water in Norway – where that water nearly killed four of us. The big difference this time was I had some naval stoker shouting and balling in my ear the whole time.

'What the fuck were you doing under my boat?' His boat? He never let up.

'What unit do you come from?'

'Where are you based?'

Then the hose started again, Paul and I getting it in turns. The force of the water was so powerful it flung me around the deck. Back into the ice bucket and the freezing water and ice. Paul was next to me; he was so close that I could hear his teeth chattering in the tank next to me.

We sensed the crew were getting tired and bored with the game now. It was only the occasional prod and slap around the head or the warm urine entering the ice bucket that I was up to my neck in! All of a sudden it was 'Endex'. The crew kicked the 50 gal drums over and we fell out over the deck. Immediately, my thoughts were 'here we go again', remember name, rank, number and date of birth. But I was lifted up and taken gently towards the mid-ship. My balaclava was removed and there was huge applause as we tried to adjust our eyes. Our enemy were now our friends and we were invited to dress and come downstairs to the officers' mess. A warm welcome awaited us and beer was shoved into our hands. They all wanted to talk about how we put up with what they had done. We moved away from the officers' mess as the young, fresh-faced officers were getting a tad excited. The senior rates' mess was a lot more friendly. The rum toddies flowed very well. Paul and I looked out over the water through the many portholes at the area where the Russian cruiser was going to be moored. Neither of us wanted to say it, but we wondered if we'd be looking out of her portholes next week, ready for a one-way voyage to Siberia.

Our business with the *Plymouth* completed, we finished our training. We had worked out lots of options, but the prime directive was to avoid getting compromised. No parade of captured SBS divers before the cameras!

Navy Intelligence wanted us to photograph the Kirov's hull as she entered harbour. This rather depended on her arriving during the day. Then we'd do a hull recce at night. No specific timings could be planned on as no one person knew exactly what time she would arrive, and the Russians weren't about to tell us.

The photographic recce was a relatively simple task. Paul's idea was for the divers to be secured on the seabed while the cruiser steamed above. It would be loud and quite an experience, but seemed the best option. During our pre-diving training Paul and I had noticed a linked chain that ran from the South Mole to the Detached Mole along the seabed at a depth of about 55 feet. Our first idea was to attach Nikon cameras with flashlights on the seabed. As the cruiser manoeuvred into the harbour they could be switched on from ashore, and bingo – mission one complete. The powerful flashlights and the remote-controlled motor drives of the camera would do the rest. Great. However, how can we guarantee that the cameras would maintain a position? What if the remote control fails? What if the cameras run out of film?

It would give us better control if we did it manually, with swimmers strapped to the line, cameras at the ready. An underwater version of the paparazzi to greet the Russian celebrity.

Lots of questions remained. What speed would the Russian cruiser enter port? There are speed restrictions within any harbour, but the cruiser would be very slow because it was entering a particularly tight area and would be required to make a sharp left-hand turn to the Detached Mole. Ocean-going tugs would help the giant cruiser into position, but they were not considered a problem as they would only help the cruiser if required or requested, once she was within the harbour. The Kirov's likely approach speed would be about two knots. That's fast for a swimmer!

What was the ship's draught? Nobody knew. All we had to work on was an estimate of 32 feet. The next question could not be answered: what would be the tidal state on her arrival? The tidal range within Gibraltar harbour is about ten feet depending on the time of year. So it looked like we had a minimum clearance of ten feet, if the ship drew as much water as we thought and the tide was out. Ten feet is not a lot to play with when you have 25,000 tons of battle cruiser passing over your head.

To test out the plan, we resorted to our old friend HMS *Plymouth*. It went well. The noise was almost unbearable, like lying down on the railway tracks and letting an express train thunder over you. The wash from the ship's propellers was fearsome. The photos came out very clearly, but we all shared the same thought: a Kirov is five times the size of the *Plymouth* . . .

The night hull recce was a different story. The initial thought was to forget the basic method of a direct compass swim; there was too much room for error with the compass and possible chance of compromise. We were told that the cruiser was going be moored on the Detached Mole. Funny really, we often wondered why the Russian cruiser was being moored at this particular point. If this were to happen to a British ship there would have been a mutiny because you don't have direct access to the quayside. No chance of a run ashore from there. The reason was that the Russian navy had a big desertion problem, so when they visited foreign ports, they usually lay off shore rather than tying up in harbour and watching half their guys go AWOL. Russian sailors could go to the most exotic ports in the world, but only look, not touch. No wonder they wanted to desert.

We submitted our plan to No. 10. We intended to hit her twice. The first time would be the photographic run while she entered the South Mole entry/exit point. We'd be strapped on to the chain that lies across the bottom, between the two arms of the mole, attached with climbing harnesses lying on our backs, in broad daylight, cameras at the ready. A ship coming right over your head is a truly awesome sight – and it feels like the screws are going to suck you up and mince you to pieces.

As the Kirov goes over, we would fire the large motor-driven Nikon cameras taking a continuous sequence of shots as the hull thunders overhead. We would be after her sonar domes, towed array and any new subsurface electronic equipment.

For our nocturnal snoop around the Kirov, we planned to wait in position, ready for the tide to come in. Swimming across the harbour, we'd get directly under the ship and photograph the hull. We'd also go up and actually feel the hull surface, two men on each side, building up a mental picture of what it looked like, what was on it. Then, as soon as we got back to base, we'd sit down and draw the shape from memory. Then it would go up to the Int. guys to work out what the various lumps and bumps contained inside her hull.

One issue haunted us. What if the Soviets sent down a team of their own divers. We knew the Russian naval Spetsnaz (special forces) teams had a fearsome reputation. Underwater, with no witnesses, there was no telling what they might do to guard the Kirov's secrets. It was impressed on us over and over again that our top priority was not to get caught. We were very competent and highly confident underwater. But our Russian counterparts are certainly a match for us: just as well-trained and just as fit. If it came down to kill or be killed, who would have the edge?

Whump, whump, whump . . . I could hear the slow beat of the Kirov's screws as she lined up on the narrow entrance between the moles of Gibraltar harbour. Ivan was coming to pay us a nice little courtesy visit. And lying on my back, strapped to the wire by my climbing harness, I was going to spy on his bottom . . .

A gigantic shadow loomed over us. The noise went right through me. The hull was so broad, it blotted out the light. Had we got our sums correct or were we going to be squashed into the harbour mud? The Kirov travelled over us with agonising slowness, the thunder of her screws now getting unbearable. We squeezed the pressel switches in our hands. The flashes lit up the bottom of the hull. The ship was moving much less than 4 knots, the usual speed that ships enter Gib harbour. I think that even the tug captains were briefed to slow her down so that we could have a good look at her bottom. On and on it went. The turbulence was incredible even at this slow speed; it rocked us left and right as we tried to lie still, attempting to judge when to take the pictures and when to hold on to the pressel switches that controlled the lights.

It was just as well that it was daylight as the flashlights lit up the whole area, even 50 feet down. The difficult part was approaching; I kept on thinking that the props of a ship are usually tucked up along the keel. This particular ship was double propped and no one person was a hundred per cent sure how they were configured. We were about to find out. It seemed to get darker. I reached over and squeezed Paul on the arm; I received a reciprocal squeeze back; we were both still OK; now the most testing part: the stern.

The battle cruiser's giant screws were thrashing towards us, churning up the water as they came overhead. I was shaking with fear; the safety margin was going to be bloody slim. The cameras flashed in the hope we'd get something, but I had to put my free hand to my face mask to hold it on. The wake nearly yanked my mouthpiece out and I had to grip it hard in my teeth to keep it in.

At last the pressure started to ease. The water became less angry and I was able to peer up at the stern as it moved on to its mooring. I grabbed Paul's arm. He must have thought the same, as both our arms touched. We unclipped the harness, excited at what we had just achieved, just hoping that some good results had been gained. We swam to the start point which was located on the outside of the detached mole. There we would stay underwater until a signal was received that it was safe to surface. Eventually, we got the signal. The team dragged us out of the water, removed our set and ancillaries and we moved under cover. We watched the Kirov slowly being manoeuvred into position on the detached mole. Phase 1 complete, phase 2 next – 1–0 to us!

The following day was taken up with last-minute briefs, with time allowed for a bit of general relaxation. Trouble was, it was impossible to get any sleep. We kept double-checking our equipment.

The task was due to start about 0100 hours, dependent on traffic and movement around the Kirov. If there was too much activity or security around the cruiser, we planned to delay for 24 hours – and thereafter every night until she left. The crew were due to be in port for a total of seven days, so this gave us the 'fudge factor' we required. We knew that the security around the Kirov would be tight so we more or less assumed that nights two and three would be cancelled.

H hour. All the moves from Coaling Island were done in covert vehicles to prearranged water entry points hidden from the public eye. Three pairs of swimmers would undertake the recce: one pair would take the stern, one pair the bows, while Paul and I studied her amidships – the area most likely to show any evidence of hydraulic doors from which the Russians might launch divers. The other teams would be looking for the prop configuration and any unusual sonar dome configuration.

The doors opened and we were out of the vehicle, down to the shoreline in the shadows. We prepared to enter the water, the surface calm and black. The final check on fins, mask, hood, gloves, set and body line. We went on to oxygen, completed our two minutes, cleared our counter lungs of any dirty air, and entered the water. We moved to about ten feet below the surface of the sea wall. We carried our compass board and constantly checked our direction just in case, so we didn't need to pop up to the surface to check our bearings, risking compromise by some sharp-eyed lookout. The total distance was approximately 1,800 yards. I swam on Paul's left, just above his left-hand shoulder. We kept at a depth of between 32 and 52 feet, going down to between 52 and 53 feet as we reached the Kirov. The phosphorescence from our bodies was very pretty, but alarmingly obvious, so we had to slow down.

It was a long swim. We had lots of time to think about the implications of being caught: Soviet interrogation procedures would make our experience on the *Plymouth* look like light entertainment. Never mind the political implications of letting our country down, we doubted we'd see our families again. (They might see us, I supposed, if they televised a show trial of the western spies.)

The noise of the great ship was all around us now. Under normal training circumstances the ship would be shut down, her props be locked in the water, intakes switched off along with anything else that might be hazardous to divers. This was different. The Kirov was a fully operational Russian battle cruiser with no reason to make things safe for us. Rather the opposite in fact. All her intakes would be working, her props could turn at any time – all standard anti-diver procedure in wartime.

The first leg went with no problems . . . second leg . . . third leg . . . and here we were. A quick check of the set: no leaks on the buddy line that we were about to depend on. The water cleared to reveal the silhouette of the Kirov. God, she was a massive beast!

We moved to the keel of the cruiser, starting our recce on the starboard side, nearest to the detached mole. Our plan was to move to the portside, complete the recce, then swim to safety. We swam deeper and deeper into the apparently bottomless harbour until we made our first physical touch with the Kirov. We immediately went into our routine. No time to hang about. We ran our hands over every crack, hole or bump we could find. Using our 13-foot buddy line we measured everything, storing the information in our heads as we'd practised.

We moved all the way down the portside until the silhouettes of the long prop shafts could be clearly seen. It was at this point that we had the fright of our lives. A small boat had been patrolling the area around the cruiser. We heard the tell-tale splash of swimmers entering the water.

They were astern of us, how far it was difficult to tell. We pressed ourselves against the ship's hull so we wouldn't be silhouetted. We watched two Soviet divers swim along the prop shafts, but they did not come up as far as us. We dared not move. My hand clasped the knife strapped to my thigh.

We waited for what seemed like hours, but we later estimated at six minutes. It must have been a routine check of the Kirov's propellers and propeller shafts. But why didn't they check along any further? In a roundabout way this suggested there was nothing of importance from amidships to the bows. On the other hand we did find a number of massive holes, over 13 feet square. Did this indicate anything peculiar about the cruiser? We could not answer that question as we were not about to commit ourselves into these black holes, especially with enemy divers in the water with us.

The Russian divers swam to the surface. After a final check we swam deep and away from the Kirov. The further we swam, the more elated we became. The swim was virtually over: we had completed a task against our biggest enemy. The phosphorescence flowed over our shoulders and around our bodies the faster we swam, but we didn't mind now. We got out of the water, tugged off our fins and masks and jumped into the van. Back to base.

We spent the next five hours drawing, describing and analysing the information. The report had to be on the plane to London the following day. Although there was no substantial proof of the ship conducting counter-espionage operations, the pictures and the results of the hull recce were a resounding success. The euphoria of it all made you want to shout from the rooftops and run up the submariners' traditional victory flag, the skull and crossbones. Two–nil to the SBS and above all Paul.

11 Back to the Job I Love

B Y THE TIME I CAME BACK to the SBS, after two-and-a-half years with the Det, I'd really lost touch with the squadron. Even so, they thought I was good enough to go straight on to an SC1's course, with the rate of corporal. I didn't. And I was right. I wasn't ready. So rather than fail it, I took myself off the course voluntarily. The following year, I completed it, no problem.

I joined 1 SBS, operating out of Poole. We covered the shipping side of MCT (maritime counter-terrorism). Five SBS, attached to 45 Commando, did the same thing only out of Arbroath, with special responsibility for the North Sea oil rigs. This was the first time that I met Chris Brogan, who was attached to the Service from the Australian SAS Regiment for a two year tour as an SBS rank went to his unit for two years. (This particular exchange had been conducted over a period of 25 years, so we new SASR very well.) The Australian SAS regiment have a history they are rightly proud of. Located in Perth, Western Australia, their links are closer to the SBS than their sister unit, the SAS at Hereford. Hereford's attitude towards them was to treat them with contempt and through this attitude their links have never been that close. The unit is extremely professional, which was once again proved in their last excursion on the operational scene – in Jakarta. Chris was from this unit, a powerhouse of a man, sharp as a razor, but with tremendous charm: a great personality. We were lucky enough to be associated with his professional attitude and never-ending enthusiasm for the job – specifically the job within MCT. As a team leader, I worked very closely with him. His planning and preparation was second to none. Chris and his wife Kathy went back to Australia to finish his time with the regiment. In that time Kathy gave birth to their son, Matthew. They talked of returning to UK and after a lot of thought and much paperwork they returned to England, and he joined us in the SBS.

Chris met with a tragic accident in Belize, abseiling from a helicopter on his final exercise. Typically, he'd volunteered to take the place of a fellow operative who was a bit under the weather. As he left the helicopter, his rifle got hooked on the 'nightsun' (powerful night light). The sling caught round his neck, strangling him. The crewman tried to pull him back inside, but he and his kit weighed a ton and the guy couldn't do it. Neither could the pilot put the helicopter down as this was an exercise with live ammunition and there were rounds going down all over the place. We'll never be sure of the exact sequence of events, but the long and the short is that they cut the rope: Chris fell like a stone from about 60 feet. He had to be resuscitated on the ground several times before he was flown to the Miami Medical Centre.

Poole was informed of the accident. Our RSM had to go to break the horrific news to Kathy. She flew out with Matthew and a close friend the next day. Kathy had to wait days before the full extent of his injuries became clear. Chris lay there in a coma with tubes everywhere you could put one. He was paralysed for sure, and the doctors couldn't tell if he had brain damage as well.

Kathy had to watch her husband fight for his life. Chris was in and out of the operating theatre for months as they fixed his back. Kathy's mental strength during this dreadful time was inspirational, especially as she had no help from

any professional body at that time or any time after. After
many weeks Chris was well enough to be flown home to
Oddstock, Salisbury for rehab. This was an equally stress-
ful time for the family. Kathy was with him day and night,
giving him the love and support that he needed. But no one
ever thought how she was through all of this, everyone was
focused on her husband. Kathy had to look after a
four-year-old boy at the same time, and get him to
understand that his dad would be in a wheelchair from
now on.

It is a credit to both Kathy and Chris that their son is a
terrific young man and as good a sportsman as his dad.
Chris was determined that his injuries would not put a stop
to his sporting career and went on to compete at international level for Australia as a triathlete and swimmer. I
admired his guts and determination to succeed when most
people in his position would have given up. That was
evident in the hospital where I saw so many others
struggling to come to terms with disability. Of course he
had days when he lost his temper. He had every right to
do so, but Chris never once tried to blame anyone else for
his accident.

The Service gave him some support but he was treated
pretty disgracefully. Here was a man of the highest calibre,
who had given up an extremely promising career in the
Australian SAS to join the SBS. Yet today, he gets more
support from the Australians than he does from us. Our
system let him down big time. And it left many of us
wondering what would happen to us if we were crippled
during an exercise. Would they wash their hands of us after
all we'd done for them? It remains a black spot in my
otherwise highly rewarding years in SBS. A few senior
people of the Service at that time should take a long hard
look at themselves. Vital evidence was mysteriously mis-
placed, evidence that would have assured the Brogans of a
fairer conclusion in the inquiry. I saw so-called friends put
their own careers ahead of the truth and justice to an

injured comrade. Only a couple of guys that he used to work with before the accident ever bother to contact him. Some invite him to a top table when one of them is leaving the Service, but in between those piss-ups? Not a word. Chris deserved more than that from blokes he gave his heart and soul to.

I felt a bit disillusioned at first. 'Normal' SF work seemed dull after the intense pressure of undercover work. We even had weekends off! But I got back into it: we were training for some of the most dangerous operations in the military world. We practised fast-roping from helicopters on to a ship moored in Portland harbour. Once on board, we'd fight from deck to deck into the bowels of the darkened ship: CQB (close quarter battle) in this environment is an adrenalin rush every time.

It hadn't happened yet, and when it did it wasn't a British ship, but we had to be ready. The fear was that a terrorist organisation would have a go at a cruise ship. The most obvious and spectacular target would be the *QE2*: 75,000 tons of luxury accommodation packed with wealthy citizens and with a British flag on the stern. Perfect for the IRA or any number of Middle Eastern types who wanted a pop at the Brits. Concern for the *QE2* was so great that the Navy posted a permanent four-man detachment on board.

It was a very sought-after posting. If you thought the rig assaults were James Bond stuff, this was even closer. It isn't often that you get issued with a dinner jacket in the Special Forces.

I had a first-class cabin and a string of bikini-clad beauties to get through before Mr Big broke down the door and gave it, 'So, Mr Bond.' Not really. I had to share a cabin with the three other guys on the team.

Consorting with female passengers was considered un-professional and was frowned upon. But we were all young and fit, if not tall, dark and handsome. The ship was packed with newly single women of a certain age, ready to blow their divorce settlement and anything else that came up, so to speak. It didn't help that we wore our pistols in crotch holsters; unlike Bond with his slimline Beretta under his jacket, we packed big fat 9 mm Brownings down our trousers. Hello, boys!

You could score like a bandit but, for the record, I lived like a monk. The only new trick I learned was to wield a lobster pick. Nice change from an ice pick.

If it came down to it, the *QE2* would be a particularly difficult ship for an SBS assault team to board. The ship is enormous: umpteen decks high and she's capable of over 30 knots. Still, we had to find a way of doing it. And make sure it would work.

We pitched up at a cavernous hangar, obviously unused for years and years, the sort you see a lot of in *The X Files*. There were no alien space craft and, worse, no sign of Scully. All sorts of agencies turned up and occupied its own allocated space: 22 SAS, SBS, police, signallers, bomb squad, logistical staff, liaison officers and representatives of the ship owners. When we rehearsed for a full-scale assault on the *QE2* the headcount was considerable. Before we knew it, the place was a bustling beehive of activity.

To us it was known as a 'mounting base' (MB), our terminology for a secure location that allows the complete force of an operational team to store its equipment, plan, prepare and – to an extent – rehearse for an operation. An MB for the team(s)/squadron is selected by the head-quarters. The operations cell, the commanding officer Special Forces and other members required by the OC move to the MB, prior to the move of the teams/squadrons/troops. While the OC liaises with the police, the ops team survey the MB area to ensure that it meets our requirements. If the MB is not suitable, we will ask for permission to recce and select another more suitable one. We need to be as close to the target as the police will allow and still ensure security from terrorists, the public and, above all, the news media. It has to be large enough to house personnel and equipment: hangar space with joining offices, toilets, wash areas. Vehicles and RIBs can be housed inside but we need space for helicopters and usually a landing strip for a C-130 Hercules nearby.

We need rehearsal areas, as similar as possible to the target building, ship or whatever facility we're going for. An additional hangar near the MB area may be used to establish a mock-up of the target area.

Once established in the MB, we prepared our (ER) emergency response first, just in case the situation demands an immediate reaction rather than the planned operation we'd prefer. If the situation deteriorates rapidly, e.g. the terrorists begin killing hostages or threaten to cause material damage in a short space of time, the ER will be stood-to and actioned if required. We don't like having to improvise in such circumstances, so a good ER plan must be thought out before you get down to planning a deliberate resolution of the situation.

Once the deliberate plan has been properly prepared and approved it may be implemented whenever the authorities give the word. Security of the MB is of paramount importance. As with any operation that we are involved in, everyone has to remain within the MB and not wander about outside. Everyone using the MB must be briefed on this fact by SSM. It is emphasised to all members that the media will be attempting to gain as much information on CT/MCT units and their possible reaction to the terrorist situation. These may include over-flights of the area. It is a sad truth that the media have no sense or regard for security. They just want to get a good story and bollocks to everyone else. Hence the compromise of British troops on active service in the Falklands by our 'own' television and radio network.

We have to assume that the terrorists will listen to radio and watch the telly to assess their impact on the public. There is also the possibility of a well-planned terrorist operation employing 'sleepers' to check areas around the target for actions of the security force. You can't trust anyone in this business.

We needed the space of a giant hangar because we drew out the decks in chalk on the floor – sometimes five or six of them, all done to scale for us to rehearse our actions on the real ship. For many liners, we managed to find a sistership and exercise on something close to the real thing. I never did find out if the *QE2* had a sistership; presumably not.

So how do you storm a liner while she's under way at sea? How do you separate a ship and its passengers from a gang of determined terrorists and still have an SBS team left standing?

From the moment any terrorist group knows the assault phase is on, the hostages' lives are in terrible danger. A sixteen-man assault team boards a C-130 Hercules with rigid inflatables; the men have on steerable parachutes, the inflatables are dropped from the plane with their engines fitted and warmed up. Weapons and kit are strapped inside. The ramp comes down, an icy draught roars through the fuselage and the boats are shoved into the night. We follow them from 1200 feet, steering our 'chutes to splash down as close to the boats as we can. The guys clamber in one by one. Quick head count. Break out the weapons and kit as we pop the clutch and go.

Far above, invisible to the naked eye, an RAF Nimrod co-ordinates the assault, its powerful surface search radar monitoring not just the enormous bulk of the ocean liner, but the tiny dots that represent our boats.

Our start point is about 50 miles out, but we close the distance fast. Five RIBs cut through the water in formation. The gigantic ship appears on the horizon, her white stern light shining brightly. Wake-up time, guys. The rhythm of the RIBs through the water, combined with lack of sleep and all that sea air has left some rather dozy commandos, doing the noddy-dog for the last few miles. Now everyone is wide awake. The long yellow poles are prepared, one man at the stern of each boat ready with the harness.

The boats split into two groups of two, which approach the target vessel from different angles. The other boats remain in reserve just in case anyone falls into the water, or an assault boat breaks down. We move in for the kill.

The final dash is the most difficult. The RIBs are flat out now, engines screaming as the throttle is wide open to close the distance. But you can barely hear them above the roar of water. We reach the *QE2*.

We are at our most vulnerable now. Poles and plummets ready, HKs point skywards. No sign of anyone on deck, but it's too dark to see to the top of the deck. The great ship looms over us. We have 50 feet to climb, while the RIB stays exactly on station. This calls for fine judgement by the Royal Marine coxswains, the best in the business.

Spray cascades over us. The boat yaws violently as we ride the bore between the towering stern-waves. She must be close to full speed: more than 30 knots, but the RIBs gain steadily, the hull thumping into the sea as we close. In the swaying assault craft it is difficult to take more than a rough aim, but the rocket-powered grapnels soar up to catch on the railings nearly 50 feet above our heads; the yellow pole wavers far above our heads; we hook on and the ladders are hauled up to lock into the top of the line. It's a long way to climb a ladder when you're weighed down with guns and ammo. The rope and ladder wriggle like a snake, swinging around with the motion of the ship and the wind. The other option is to use magnets on your hands and feet – the 'human fly' routine – but it takes for ever. You eventually get on the ladder and start the long climb up. Your arms burn as you hang on for dear life, trying not to see the white water surging below and the boats fighting to stay on station.

Up and up, wondering if this will ever end. Somewhere up there is the wooden lip that encircles the chain deck. No one's seen us yet. Lungs begging for air and arms nearly wrenched from their sockets, you haul yourself gasping on to the deck, knowing that this could be the opening second of a gun battle. We scan the deck over the sights of our HKs. Now there's a third man up. The rest of the team quickly establish themselves in all round defence, just in case we've been compromised. No sign of life. We signal 'dry feet' to the Nimrod. Let's go to war: the team pads silently across the deck, heading for the priority targets. In the pounding sea below, the RIBs peel off and move to a holding area some distance behind the target. Eighteen minutes to H hour.

We practised attacking the *QE2* every time she went in for a refit. The great liner would dock at Southampton and discharge 2,000 or more passengers before heading off to Hamburg where the work would be carried out. On the short run to Germany she'd still have 400 or so people on board – mostly crew. Among them, unknown to anyone, would be a team of 'terrorists' including at least one policewoman, waiting for the exercise to begin. Of the ship's personnel, only the captain knew what was going to happen – and when. The first thing everyone else knew was when five giant CH-47 Chinooks appeared above the ship, spilling ropes from their cargo doors. Down came the Men in Black brandishing machine guns.

Flying one of these monstrous helicopters between the ship's masts in the dead of night – with no navigation lights – is the supreme test for the pilots of 7 (SF) Squadron. They really do have just inches to spare. The pilot has to maintain the exact same forward speed as the ship while keeping his blades within the tiny clearance available. One mistake and everybody dies.

Every time we did this the crew ran about in panic. While some cottoned on quickly, most hadn't a clue what was going on and thought it was the real thing. One moment they were steaming along, thinking ahead to a run ashore in Hamburg; the next minute the place is crawling with commandos – bug-eyed monsters in S10 respirators.

Once we'd dived down from the Chinooks or climbed up from the RIBs, it was time to play 'hunt the terrorist'. Just as in real incidents, once the gunfight went against them, the terrorists would often dump their weapons and try to hide among the passengers. We had to switch from CQB mode – flash-bang ... double-tap ... double-tap – to reassuring civilians and weeding out the bad guys. And girls. Concealed 'terrorists' were apt to pop out from the woodwork to liven things up. In coastal waters or in harbour, civil authorities would be able to take charge quite quickly. We'd be withdrawn and the police would take over, but if this went down in mid-Atlantic, we'd have to sift through the passengers to separate the 'X-rays' (terrorists) from the 'Yankees' (civvies). We used the liner's enormous restaurants as holding areas while we proceeded with the interrogations. We ran the show until we made landfall or the police are flown in to take the handover from us. Some of the exercises do have a detrimental effect on even the acting terrorists.

With the post operation procedure complete we found that one of the policewomen was suffering the real thing: a week of being questioned by the negotiators took its toll and she was off work for some seven months. Post operational routine is just as important to get right as the actual attack.

Cunard appreciate our efforts and support us in what we do. It has been a long-standing relationship and one that will I'm sure continue. After completing the exercise, we are assigned apartments on board and the complete team is invited to one of the main restaurants for a monstrous feast. There are some compensations.

12 Down South – The Falklands War

I DIDN'T EVEN THINK ABOUT what I was leaving. My wife, my son, my daughters, my home, it all paled into insignificance. This was the job, the whole job, and nothing but the job. When the call came to report to Faslane, my only thought was, 'We'll be going down there by submarine, then.' And that was all. I started getting my kit together. I was whistling and chirruping to myself as I packed. I thought I was alone until I heard Fiona sobbing in the corner.

'What's the matter, honey?'

'You like doing this, don't you?' She glared at me.

What had I done?

'You don't care at all that you might get killed,' she snapped. 'This is a war!'

'It's my job,' I said. 'It's what I've trained for. It's who I am.' Then I softened up a bit, realising how very upset she was. 'It's OK,' I said blithely, 'I'll be coming home on the return trip.' Nothing like absurd optimism. I was an operational team leader in the SBS, this was what I'd trained for and I was going to do it. Family or no family. I admit I'm a bit ashamed of my attitude now, but in 1982 no one was going to stop me getting to the Falklands.

We reported aboard the nuclear attack submarine HMS *Conqueror* at 0730 on 11 March 1982. We left Poole with every bit of operational kit we had in the lockers: weapons, ammunition, demolitions gear, diving sets, boats, canoes, engines, Bergens, belt-orders, personal kit, the works. It needed a nuclear submarine just to get the stuff on.

Faslane is notoriously difficult to get into. On this day, however, we drove straight in through open gates. I was astonished. But we were individually checked for identity once alongside HMS *Conqueror* on the quay. They were loading live Mark 24 guided torpedoes and the old un-guided Mark 8s, the long slim tubes gleaming dully in the morning light. But they left the bottom torpedo racks in the fore ends empty, and that's where we put our stores.

We sailed for the South Atlantic 24 hours later. It was the last time we would see daylight for 21 days.

The Falkland Islands – or Las Malvinas – in the South Atlantic had been the focus of dispute for decades. In early 1982 the British government announced that our minimal naval force on the islands would be reduced. It was the chance that Argentina had been waiting for and Argentine forces invaded in April. The paying off of HMS *Endurance*, the decision that triggered the crisis, was cancelled. Her captain had read the situation correctly, reporting the massive increase in signals traffic over the months leading to the invasion, but he was ignored by the Foreign Office. Substantial cuts in Britain's armed forces, especially the Navy, were suspended at the last minute.

By this time we were half way down the Atlantic, moving at 30 knots at 600 feet. This was a precautionary move in case it came to a shooting war and the Prime Minister – Margaret Thatcher – decided to respond with military power. The Argentine garrison was strong when looked at on paper. It amounted to some 10,000 men with 105 mm light howitzers, 155 mm gun/howitzers, and AML 90 armoured cars. They were reported to have 25 heavy and medium helicopters, 15 Pucara ground attack aircraft on the islands themselves. On the mainland they had a powerful air force with some 75 modern fighter-bombers and 5 Super Etendards armed with Exocet anti-ship missiles. It was not going to a pushover. However, the two main constraints were distance and time. Our high command's main aim was to get there with sufficient forces to complete the reoccupation in six to seven weeks before the winter set in.

The biggest threat to the British task force was from the Exocet missiles which could be launched from the mainland, from the Falklands themselves or from enemy warships. On paper, they had air superiority. In comparison, the good old British Navy was geared up for NATO deployments, focused mainly on ASW (anti-submarine warfare) in support of US Navy task forces. Scandalously, the Royal Navy had no airborne early-warning aircraft. Systems had to be improvised during the war. Meanwhile, the answer was simple. Tell the Special Forces to get to the Argentine coast in HMS O Boat and get their asses on shore. Phone if you see any airplanes heading out to sea! Other SF teams would land on the Falklands to check for Exocet positions there too.

What did we have to go down south? Type 42 destroyers, *Coventry*, *Glasgow*, *Sheffield* and *Glamorgan*; the frigates were *Arrow*, *Yarmouth*, *Alacrity*, *Ardent*, *Antelope* (my favourite frigate – which I saw sink) and type 22 frigates, *Brilliant* and *Broadsword*; aircraft carriers, *Hermes* and *Invincible* (with a total of 34 Sea Harriers); fleet auxiliaries *Olmeda*, *Resource*, *Fort Austin* and landing ships *Sir Galahad* and *Sir Tristram*. It amounted to an impressive array of shipping power, which in effect was self-sufficient from a distance of 5,000 miles. It was very obvious to the man on the ground that Thatcher did not want to take any prisoners.

The transit to the Falklands was a remarkable experience. The conditions were cramped even though one third of the *Conqueror*'s crew had been left behind to accommodate us. We got on well with the crew. You would expect the days to be long and boring, stuck in a giant cigar tube, but there was always something to be done. The boss and team sergeant major were of the attitude that we should check and double-check everything, from the water bottles in our belt order to the huge amount of explosives that we'd brought with us. We went over our standard operating procedures until they were second nature – well, on paper anyway!

This routine was good for the first week or so. It was then that Captain Chris Wesford-Brown, a very sound leader of men, did a great job of maintaining morale through his witty and frank daily briefings. (Some would say that is what he gets paid for. I disagree; having been in and out of submarines for over twenty years, I have seen some crews with very low morale, usually due to their captain and officers.) This man was different: the best I have come across in the submarine world. He decided that the best thing for all concerned was that we should be taken into the watch system of the boat. This worked out well, but most of all, he kept the team interested and happy – by letting us drive the submarine! I became quite an expert at working the hydroplanes that change the submarine's angle of attack in the water, and hence its depth.

We occasionally came up from 600 feet to speak to London. At periscope depth (PD), the antenna can project above the waves for radio communication. Only once did we contemplate surfacing, when the communications antenna got damaged. We wallowed just below the surface while the head shed weighed the pros and cons of bringing her fully to the surface and risk being compromised. It took them over six hours to decide, by which time we all suffered from sea sickness.

One of my team members was not a happy little bear. I thought he was beginning to crack up with fear doing a live operational task. All he kept saying was 'Do you think this is really going to happen?' bearing in mind the country had yet to declare its intentions fully. All I could do was reassure him that it was and that he should just get on with the job in hand. It all seemed a bit pointless at that time. He was an extremely fit man who was totally against smoking. The next day he had purchased some roll-ups and started smoking. He became known as the 'low moral fibre' case and got worse as we got closer to our op area.

I passed the time reading Winston Churchill's *The Finest Hour*. I'm a great fan of his. If you're on the way to fight a war, there's nothing like Churchill for getting you in the right mood. Keeping the mind fit was one thing, but physical fitness presented a bigger problem. The sub was cramped at the best of times and now it was crammed to capacity with live ammo. We did endless pull-ups on the bulkheads.

Nuclear submarines are the Royal Navy's capital ships. They're huge, terminally expensive, and the Navy's determined not to lose them. When we surfaced, there were no fewer than five frigates and destroyers around us. Further away, out of sight, were additional warships with their radar teams scanning the horizon for the first sign of an enemy presence. We even had an O-class submarine out there, creeping along on its electric motor so as to be as quiet as the church mouse. Its mission was to listen for enemy subs: we knew the Argentines had a couple of new German-built diesel electric boats. If one of them got close enough for a shot, he'd be the toast of Buenos Aires. (He'd also be toast, period, but that wouldn't be much compensation if we ended up on the bottom of the Atlantic.) Air cover was provided by Sea Harriers from the carrier *Invincible*.

Having busted a gut to get there, we ended up staying on board another fortnight. You could feel the frustration in the air. We were desperate to get ashore and do the job. At last, they sent us on to South Georgia. The next day, *Conqueror* put two of her Mk 8 21-inch torpedoes into the Argentine cruiser *General Belgrano*. World War II weapons to sink a World War II warship. (The sexy new guided torpedoes were apparently regarded as rather unreliable. Mk 8s had no guidance, but carried a whopping great warhead.) The captain had been on the satellite link for some time, pleading for a change in the rules of engagement. This was granted just in time to take out the second biggest ship in the enemy fleet. *Conqueror* thus did exactly what she had been built for, before she acted as a taxi for the SBS. The enemy fleet fled back to port and stayed there for the rest of the war. The captain famously returned to Faslane, flying the skull and crossbones – a traditional gesture by RN submariners during the world wars.

I was going to say that South Georgia is in the middle of nowhere. This is true, but more relevant is the fact that it lies on the edge of Antarctica. The island is a mountain, shrouded in a glacier, with a few inlets where whaling ships used to put in. The Argentines had overrun a tiny Royal Marine detachment here, but not before Lieutenant Mills and his team took out one of their helicopters and engaged a corvette with the 84 mm Carl Gustav anti-tank rocket. They holed it below the waterline too.

We did extensive planning on retaking South Georgia. We moved to county class guided-missile destroyer HMS *Glamorgan*, then RFA *Fort Austin* and sailed south with the task force. There was a lot of inter-service in-fighting about who should go in and do it. D Squadron, SAS, was on board *Fort Austin* with us. We'd done the appreciation on the operation, on the *Conqueror*. They'd also done one. We both realised neither the SBS nor the SAS were qualified for this task. In fact, what the job really needed was the Royal Marines Mountain Leaders platoon, but who are we to suggest such common sense solutions! Even so, the SAS (affectionately known to the captain of *Fort Austin* as the 'Friday afternoon soldiers') managed to bulldoze their way into a situation they were not qualified to deal with. It was a giant SAS cock-up on Fortuna Glacier, a near catastrophe.

This glacier is gigantic, a half-mile thick sheet of ice rolling off the land and right down into the sea. Great chunks of it break off as you watch, cracking and groaning into the water. Getting up it was going to take everything a specialist team of ice-climbing troops like the MLs had. We thought of landing on it from the air, but in the unpredictable South Atlantic weather, the idea was suicidal. We were very nearly proved right.

Letting the regiment have its head was a bad command decision. Twenty-two SAS weren't invited to go down to the Falklands, they just turned up on the quayside. Few, if any of them, except for Lofty A, had any experience at all of working on glaciers. After seeing them in action, and losing his helicopter, *Fort Austin*'s captain renamed the SAS 'Friday and Saturday afternoon soldiers', because he was so fed up with what he saw as their unprofessional 'we can do anything, because we're the experts on everything' attitude.

Once the SAS had covered themselves and everyone else in confusion, 8 SBS went in by Wasp helicopter to Hounds Bay, and later by boat when flying became impossible. They negotiated the ice-field in appalling weather and RV'ed with the remainder. Even at this stage the weather was horrendous. The team had tried in vain to cross Nordenskjold Glacier in the only one serviceable Gemini. It was impossible that particular day, so they decided to try again the next. At this time there was no communications to the *Antrim* as the fleet had pulled fifteen miles away from South Georgia due to the presence of an Argentine submarine – the *Santa Fe*. The SBS teams had been left behind. In the end, due to weather conditions, the team was eventually withdrawn. With hindsight, the Argentinians had suspected that something was up and proceeded to find out. Grytviken was eventually retaken with a combined team of SBS/SAS and Marines. It was a success and Prime Minister Margaret Thatcher commanded the nation, 'Rejoice, rejoice, thank heavens for our Marines.' If she only knew how some of Britain's finest nearly died.

Most military units came out of the Falklands War with their standing enhanced. The reverse was true of 22 SAS. In many senior people's eyes they were aloof, cut off from reality. They had an attitude problem with other personnel who were working exceptionally hard to help us as SF troops – good old 'jack'. Certain elements of their regiment continued to let them down later in the campaign on the main islands with so-called forays into enemy territory, reporting tasks that they never completed or even started. They would be communicating from their drop-off point – where they remained. Thankfully these prats were returned to their units when they got back to the UK: there is no place for such people within SF. To be blunt, they were putting the lives of the Paras and Marines at risk.

From South Georgia we moved on HMS *Plymouth* ready for the assault on East Falkland. The task force wanted no nasty surprises when the main force came ashore, so SBS teams reconnoitred every possible site. The Directing Staff (DS) solution was to exploit the wide beaches right next to Port Stanley, pile in there and end the campaign at a stroke. And that's the problem with obvious answers: it's exactly what the enemy expected. Their marines were trained by the US Marine Corps and the garrison had been busy sowing mines along the beach, and zeroing in their artillery.

So our interest turned to the other side of the island. As we discovered, there were Argentine forces there, but their defences were not as formidable. They clearly did not anticipate that we would be mad enough to land the 'wrong' side of East Falkland. It committed us to a long slog on foot across the god-awful terrain, on the very eve of the Antarctic winter. We were tasked to clear a number of areas on East Falkland of Argentine troops: Teal Inlet, Fanning Head, San Carlos Bay. Our initial entry was from *Plymouth* via the ship's Lynx helicopter. It was to be a two-night, 25-mile walk to the western side of Teal Inlet. I remember this particular insertion well. It was the first time that we were going to be in the field in excess of ten days. Our five-man patrol carried enough weapons and ammo to take on the whole Argentine army. My 'low moral fibre' man kept wishing the war would end before we went in. He was disappointed.

There were four teams to be inserted that night, one at a time. The Lynx could not make a landing, so we had to judge the height and jump. We expected a jump of four to six feet – not too bad. The Lynx reached our drop-off point and began hovering. I looked down. I couldn't see a thing. So I pushed my Bergen out and it bounced on what I assumed was long grass. No problem. I jumped.

I fell for far too long. Where the hell was the ground? Should I adopt free-fall position? I was in the air for a fucking long time before I too bounced on the thankfully long grass. I looked up and watched the second man clamber out and jump: yes, definitely at least 30 feet! All the team landed safely except that my 'low moral fibre' case appeared without his M16. I was livid, so annoyed that I dragged my 66 mm from the side of my pack and shoved it in his hands. 'There,' I shouted. 'That's your personal weapon for the duration of the op!' The Lynx flew back and landed on *Plymouth* with his M16 tangled in its skids.

We secured the area for the other teams to come in. Tony brought in the M16 for my LMF case. This particular op went without any further mishaps. Our major problem was finding good areas for our OPs and LUPs. The highlight of the op was the RV with the other teams. It was great to see them after twelve days and count everybody in, as it were. The EXFIL (exfiltration) by the Sea King was a stunning example of NVG flying. We had recced the LS and were expecting the helicopter. It went like clockwork: the lights went on at minus 2 minutes on the small plateau at the end of a deep, long running re-entrant. Right on H hour this thing appeared from nowhere. We did not even hear the helicopter approach. It slid up over on to the landing site, the massive frame of the Sea King suddenly plonking itself down right on the money. Outstanding. No sooner than it was on the deck, the sixteen-man team was in and the Sea King moved a few feet, turned and slid down the re-entrant and away. I briefed the command that the western side of Teal Inlet Water was free and available for a landing.

The Falklands are worse than a Welsh bog on a bad day – bleak, deserted, wet, horribly exposed, and with occasional nasty surprises around the corner. One of our biggest problems was getting good cover during the daytime, both from the air and from the ground. Usually, we'd start looking for an LUP at about four in the morning, and it would take us until seven o'clock to establish one.

The bald, peaty terrain meant we had to dig down about one and a half feet into the ground, carefully strip off the top layer of turf, put this and the dirt on to a poncho, conceal that, put a camouflage net over our shallow grave, and get into it. Then we'd lie there all day, like little fat larvae, waiting to hatch. The waiting was the hard part. The slight depression meant that we at least got a little shelter from the perishing wind, but if it rained (which it did all the time) we got soaked, and we stayed soaked.

At last light we lifted the poncho back to the site and replaced everything exactly as it had been. I was glad of my Norwegian ski march boots, which were waterproof up to the calf, but, being made of leather with a flexible top, they allowed the accumulated sweat to escape. It was immediately obvious in the Falklands that trench foot was going to be a big problem. I also had lightweight windproof trousers and top over an arctic T-shirt and a shirt that I would take off and bag in the Bergen once we'd got going.

All across the islands, two- and four-man OPs were in position, monitoring the Argentine garrison, observing their routine, deployment, state of discipline, etc. On the evening of the landings, we cleared the Argentine observation posts on the high ground around Fitzroy Settlement. Our aim was to give the Paras and the Marines a clear run in for their assault. We used two borrowed motorbikes to scout on our flanks as we contoured up.

Our first contact was with an Argentine OP: a tent with a low wall of stones piled around it. We crept forward, but there was very little cover out there. If they were watching – and they should have been, it's what they were there for – we would be spotted and brought under fire any moment. We kept going, always with 'one foot on the ground' i.e. guys covering the enemy position with their weapons while others crept forward.

We got to within 600 feet. How could they have let us get this close. Was this a trap? Time to find out! I gave the order to open fire. The GPMG gave it, *bang, bang, bang, click* . . .

'Stoppage!' screamed the gunner.

'Shit, shit, shit!'

'Give us the '66,' I yelled.

I yanked back the tube to arm it, pressed down the detent button and slammed it into my shoulder. Quick squint through the plastic sight and I squeezed the rubber tit. I couldn't help blinking as the rocket shot out of the tube, but instead of seeing the enemy OP vanish in a cloud of smoke and flame, I watched the rocket slither across the ground 300 feet in front of us.

'Fuck! It's a dud!' I didn't know whether to laugh or cry.

At this point the language deteriorated. With our machine gun jammed and anti-tank rocket useless, we opened up with our rifles instead. To our amazement, that was all it needed. As our first rounds tore into the position, the Argentines inside bolted. Unfortunately for them, they ran into the cut-off groups I'd positioned at the re-entrants either side. Now they were the rats in the trap.

They never fired a shot. Just stuck their hands in the air. We skirmished forward to take a look at our prisoners. They were young conscripts. All had shit themselves with fear. Their feet were torn to pieces: they'd been asleep with their boots off when we'd opened fire. They'd legged it over some nasty rough ground only to run into more British soldiers. We checked out their position for maps, anything of intelligence value, but it looked like the aftermath of a jumble sale.

We went up each peak like this, clearing the Argies one group after another. Then we came to the big one. This OP was a big bastard. It was three o'clock in the morning. There were five of us, and twelve of them. We had a lieutenant with us, from the Royal Artillery, a gunfire support specialist. Standing off at sea, we had a frigate with a nice big fat 4.5-inch gun on her fo'c'sle. The enemy were in the dip of a low saddle on the peak ahead of us. A quick, whispered command on the comms, and the gunner had called in 'illume' – illuminating, or 'star' shell. A sudden *whooshhh* over our heads and they burst right over the Argentine OP, bathing everything in an eerie light, turning the countryside a menacing electric yellow-white.

The enemy's first reaction was to panic. Then some of them began firing back. The sight of tracer floating up towards you concentrates the mind wonderfully, especially on the thought that for every red round you can see there are four more in between that you can't.

Our gunner ordered high explosive, tested the fall of shot and adjusted it, giving a new grid reference for the unseen ship offshore to fire on. Within a minute, the shells were bursting square on the Argentine position. Even at 700 yards, the explosion of a 4.5-inch shell makes the ground shake under your feet. The flash is blinding in the night. We watched the scene through NVGs, a war movie in green and white. Stones and bits of tent flew up into the air. That was how to clear an OP in style.

The weather was unbelievably filthy. We were thousands of miles from the UK. The Argentines had put a massive number of troops on the islands, far more than we could match, man for man. I don't think they ever believed for one minute that we'd come down there and push them off. They still didn't believe it, even when it was happening to them. That's reality for you. But we didn't care about the weather, we'd trained for it in places that could be even worse, like Norway. We were inured to it; they weren't. Almost every position we overran was in a sorry state, rubbish everywhere. They seemed to have crapped wherever they pleased, another sign of poor discipline, so you had to be careful where you took cover. Only one enemy position gave pause for thought. A group of Royal Marines were pinned down by a sniper for a bit. When they reached the OP, they found it neat and tidy, maps and papers burned, radio smashed and turned to a wrong frequency . . . all the signs that a professional outfit had evacuated in a hurry.

We were given a 'troop deliberate task' to clear Fanning Head, the gateway to San Carlos Bay. The briefing took place in a conference room on board our forward operating base (FOB), HMS *Fearless*. The mission was to clear Fanning Head for the SAS, so they could have a clear run at the Argentine airfield there. Fanning Head was being used to monitor San Carlos Bay and be a possible help as an early warning system for the aircraft on Pebble Island. Insertion was going to be by helicopter or inflatables at troop strength – four teams of four or five, plus an HQ element, and a detachment of mortars from the SAS (they do their best to get everywhere!) – and HMS *Antrim* was going to be our FOB.

The command did not tell us that we were carrying out this task for the SAS, nor that the SAS had muscled its way in to making an attack on the Pebble Island airfield; they were not as well qualified to do this as we were so we'd have been just a tiny bit annoyed about it. It was all a bit ass about face really; the SAS should have been doing this job for us so that we had a clear run into Pebble Island; then again, who are we to come up with these common sense solutions? No, I think someone probably had a choice of ops and the Pebble Island would have a bit more PR behind it!

We went into Fanning Head expecting the worst. We each had twelve 30-round magazines for our M16s; HE grenades, smoke grenades; '66' anti-armour rockets; we all had NVGs, and we wore them all the time. I was carrying an L42 sniper weapon, as well as my M16. We needed every last ounce of firepower.

We left the *Fearless* at about 2330, arriving at Fanning Head at 0230 after a sickening ride, our engines sound-proofed with special covers, which tend to overheat anywhere other than the Falklands, the ice-hole of the world. We came in three miles short of Fanning Head, stopping at about two and a half miles short of the target for a final RV to confirm our exact position, map-to-ground, and to cache our Bergens.

Now we were in battle order. We advanced up to the enemy bunkers. Underfoot was the standard Falklands tufty grass, long, sharp-edged, inclined to trip you – bloody hard going. It was a still night, with good visibility and intermittent light drizzle. The Argentines were at home. Through our NVGs, we could see they were well dug in. We lined out, all extremely excited – and extremely apprehensive. We did not know if these were special forces in front of us, or conscripts. Would they run for their lives or fight to the death?

We advanced with great caution, the groups leapfrogging each other so the enemy position was covered the whole time.

Then they saw us.

We had a Spanish-speaking Royal Marines officer with us.

'Do you want to surrender?' he shouted. He was a pretty baby blond, and he looked about sixteen years old. I think it was his extreme youth and fairness that offended them. No sooner had he finished speaking than they opened up at us with everything they had.

We all dived for cover.

'Christ!' swore the officer, crawling up to me in the dirt, an incongruous grin on his face. 'Um, they say they don't want to surrender.'

'Really?' I replied.

When their first blast died away a little, we started giving them some back. Our firepower – several '66' rockets, four GPMGs, M16s, sniper weapons with night-sights, 203s firing 40 mm grenades – was far greater than theirs, and greater than anything they'd apparently expected. No stoppages this time: the GPMGs thumped away like fury, the rockets exploding with deafening *whumps*, sending clouds of turf and stones into the air.

Inexperienced troops tend to fire high at night. Even very experienced troops may take some time to get their tracer down on to the target where it belongs. Ours started to whack home on to their dug-outs almost at once. I was laughing and chuckling like a madman, my beserker's reaction to nerves. In training, we're taught to distinguish between the *crack* and *thump* of high velocity rounds: the first is the bullet zooming over your head at supersonic speed; the second is the sound of the shot itself. The more time that elapses between the two, the further you are from the shooter. At Fanning Head the cracks and thumps were pretty indistinguishable. The Argentines blazed away and the noise was incredible.

Suddenly, I realised that the enemy's fire was slackening. Our controlled fire was taking its toll. There was a long silence from the enemy, followed by sporadic return fire. Somehow, we could tell they'd lost their balls.

We started pepper-potting forward, shouting all the time to one another, our fire-support teams keeping the enemy's heads down with the chattering GPMGs. In short rushes, we raced at the enemy, alternately sprinting across the treacherous moon grass, then hitting the deck and firing again.

We were almost close enough for grenades when they cut and ran. When we counted the bodies, they'd left twelve dead, three wounded and nine prisoners. Whether the casualty list was due to HMS *Antrim*'s 4.5-inch gun we do not know; however, if it was, some of them were killed twice! Oddly, we had no wounded. Not one of us had a scratch except one of the team needed a new Bergen. We reorganised, distributed ammunition and checked out the position. Then it was back down to San Carlos Bay, a two-night, 25-mile walk. We checked out the whole area for a possible landing, while other teams came in from the sea, examining the gradient and make-up of the beaches.

Eventually, we were able to report that the shoreline of East Falkland was clear in the area where the task force proposed to land. The fleet slipped into Falkland Sound at the dead of night and the landings were well under way before the first Argentine reaction.

We were based on *Fearless*, anchored in 'Bomb Alley'. Frankly, we were always glad to get ashore where it was safer. The Argentine air attacks were incessant. Jet aircraft skimmed the surface of the sea while machine guns hammered away at them and missiles left plumes of smoke in the air. I saw HMS *Antelope* go down, 900 yards away. One of the saddest things I've ever seen, that graceful little Type 22 sinking.

We moved to Teal Inlet to start clearing the area for the arrival of 2 Para. It was a long week; however, once complete, we eventually RV'ed with the CO of 2 Para and guided them in for their assault on Mount Longdon. They were a great bunch of professional soldiers who had a great sense of humour. The old Para/Marine stuff came out, as it does, but in a good-natured way. This was a live operation. The Paras wanted their flanks protected by the Marines and no other unit. The Marines wanted their flanks guarded by the Paras. Beneath all the banter is a bond of mutual respect. If you look at each assault on those hills, that was normally the case. They were the units that did all the hard work. It was a great feeling, being weeks or months ahead of any task force and having no support, but assaulting those hills was back to World War I. Those battalions deserved every medal that they received.

Our next task was to recce and, if possible, clear the Mount Low peninsula just to the north of Stanley. We landed in two rigid inflatables, eight of us in all, on the south side of Berkeley Sound. For once, the sea was calm, and there was no moon. Six hundred feet out from the landfall, we switched the engines off and began to paddle in. The beach we were aiming for was semi-enclosed, with two arms of high ground on either side of it. This magnified the slow swish of our paddles, making them seem very loud.

We had no intelligence of any Argentine presence actually on the shoreline, so we weren't expecting any trouble. But we got it.

We'd just got the first man in the water, with his Bergen on his back and water up to his knees, when the whole world started firing at us. There were two whole companies of Argentine troops dug in all along the landfall, and they could see us clearly against the sea. At least two hundred men, dedicated to killing us. I looked up, and saw the whole sky filled with orange tracer, long fingers of it questing and pointing at us. I'd never seen anything like it. For a second, I froze solid. Then we started panicking.

The guy already in the water slung his Bergen back in our boat, and heaved himself in over the side after it. Immediately, we grounded. The man on the engine couldn't get it started. We heard the engine on the other team's boat cough into life. The firestorm was zeroing in on us now, the bullets whipping into the water all around the boat. The gunfire sounded like a continual roll of thunder, the tracer silhouetting us starkly against the sea. We felt as if there were searchlights on us. We were sitting ducks.

We were still stuck fast.

'Get out of the boat,' someone shouted. 'Everybody out!'

I realised the force of this, and jumped into the water. Everything had slowed right down. It seemed to take me a year to get my feet over the side, another century to turn round and grab the gunwales to push. The inflatable bobbed free. We began shoving it out to sea. There was a cry from Mikey, next to me, as a 7.62 mm round hit him. Still under fire, we scrambled back into it and started paddling for our lives. The engine still wouldn't start. At last, the coxswain got it going, and we just lay flat, listening to the rounds thwacking into the fibreglass sides.

It's a curious thing about gunfire, but you often seem to need an awful lot of it to kill people. The Argentine troops had put down thousands and thousands of rounds on us, and the only casualty we had was Mikey, who was shot through the hand as he was trying to push the boat out. Why we didn't all die there in the water I shall never understand. But I understood all right how afraid I was. I'll never forget it.

Every Friday morning the Argentine general staff held a meeting in Stanley Church. Which made them a nice, fat, dependable target. Our fun little boat trip had been the first probe from the north towards Port Stanley, and it taught us that the Argentine defences on that side were formidable. Despite that, T.J. Thomas, our CO, asked me to take a four-man team, a Milan anti-tank missile, and see if it might be feasible to put a rocket in through the window of the church. This meant landing on the Stanley peninsula proper, which was to be even more infested with enemy troops. We'd have to work right under their noses. I could see it was possible; if we were lucky we'd be able to get the missile launcher set up, and probably take a shot. It is incredibly accurate and has a range of nearly 1,800 yards. All of which was fine, until it came to getting out again.

I looked at the map, looked at T.J., and remembered the firepower we'd come under on the more northern side.

'I'm willing to do it,' I said, 'but you do realise this is a one-way trip?'

After a short review with his staff, he came back to me. 'You're right,' he said, 'it's a suicide mission. We'll forget about that one, then, shall we?' I needed no encouragement to agree with him.

When we got back on board the *Fearless*, the first casualties were coming in. There were Marines with their legs and arms blown off, leaking blood from the tattered stumps. The medics had them lined up outside the sick bay, and they were going along the line sorting them out in order of priority.

'He'll live . . . He won't . . . That one's going to die, put him to one side . . . That one with the missing hand, he'll be OK, make him wait at the back.' And so on. These were blokes I knew personally, and even the ones I didn't know I still felt that strong tie of camaraderie. It was an awful, horrible sight, and anyone who thinks war is glamorous should be made to spend twenty minutes in a battlefield casualty clearing station. They'd bloody soon change their minds.

With the landings at San Carlos under way, the next priority was the ground clearance of the West Falklands. I led the first SF team to go on to West Falkland. As the Royal Marines and Paras marched on Goose Green and Stanley, we marched on to do our advanced force mission: clearing the ground ahead. We finally had clearance to approach the civil population. We landed by helicopter at about 0200 hours. The area was open, covered with lush green grass that stood some three feet high. We remained there looking and listening for a good thirty minutes. Once I was sure that no one had seen us, I decided to move out to our first contact with the locals and conduct a bit of 'hearts and minds'.

We moved cautiously across the ground until we arrived at our pre-arranged location, a small isolated farm. We patrolled further west to clear the high ground and remained in the area for two days before we made an approach to a local farmer. We established an OP in a hedgerow some 300 feet from the main farm buildings. No sign of enemy movement. No sign of any work being done around the farm. We counted two adults: one male and one female, and a small girl and boy. On the morning of the third day, we decided to move in. The plan was that myself and Phil would move forward to the front door of the farmhouse, while the remainder of the team gave us cover. We moved carefully to the door and knocked once.

A lady answered the door. She had on a long woollen skirt and a baggy jumper, plus wellington boots. She was as nervous as a cat, but the moment I spoke English, she burst into tears.

'Thank God you're here!' she cried and hugged me. Now I'd been in the field for four days and I could only guess how we all stank. (We get used to it!) Her name was Mary and, in a few moments, her husband John was down to introduce himself and the children were running all over the shop. They invited us into their cramped, but warm and comfy lounge. Yes, they said, there were Argentine troops about on West Falkland. Their patrols had visited twice and asked if they'd seen any British soldiers.

I left the remainder of the team on the ground until late afternoon while Phil and I checked the farmhouse and the surrounding buildings ensuring that there were no enemy lurking out of sight. At last light I asked Mary and John if I could bring in the remainder of the team. They were delighted and gave us a room big enough for all five of us. They celebrated our arrival by slaughtering one of their cows. It was the first solid food we had had for days. John put the massive side of meat from the cow directly into the peat oven, which was the centrepiece of the lounge. This piece of meat was as big as a car engine block and it took twelve hours to be cooked enough so that we could munch on the outer casing. We lived on beef and roast potatoes for the next three days.

We kept a watch, while giving John a hand with a few odd jobs around the farm. We struck up quite a relationship with the family, bimbling about with our first line equipment always on our bodies – including our pistols.

Our HQ eventually told us to get on with the war but, in the meantime, Mary and John sent out a signal on their radio. Travelling via the radio ham community, it informed our wives and girlfriends in the UK that we were safe and sound.

I must admit that we were quite reluctant to leave the warmth of the house and hospitality of Mary, John and their family. But off we went, into a cold and wintry evening. We established an OP on the extreme tip of a peninsula directly overlooking the eastern end of Pebble Island. Right across from us, there was a company-level unit of Argentine troops. Since we couldn't dig out a concealed position, because we were on hard rock, we built up a rock hide in a shallow concave slope, disguising it as best we could. It was impossible to defend from behind, so we put out Claymore mines to the sides and rear. The seagulls, thinking the short visible pieces of command wire were worms, kept swooping on them to gobble them up – only to find that the 'worms' were 65 feet long and attached to high explosives.

On the second night, I woke up to see a strange red light glowing in front of me. Now I saw it, now I didn't. I sat up and grabbed my M16. Then I saw what it was: one of my men had put a hole in a slightly opened matchbox, and he was smoking with his cigarette concealed inside. What he didn't realise was that while the tip of the cigarette might be covered, his face was glowing bright red every time he took a drag. If I could see it, there was every chance the enemy could. I was furious. He was putting the whole team at risk. But because we were in a tactical situation, I couldn't let rip. I decided then, though, that he'd had his last chance. He'd been an LMF (lack of moral fibre) case for some time now. When we got back to England he'd go out of SF for good. My initial thoughts were 'I'd make sure of it'. However, you do calm down as time goes by and I did mellow out by the end of the war. After all, I wanted to achieve my aim of going to a war zone with a five-man team and coming home with the complete team. Not that I had long to worry about it because suddenly we had a very different problem: an enemy 'spy ship' was shadowing the task force and we were withdrawn from the relatively quiet West Falklands to deal with it.

13 Contact Front!

T HEY'D BEEN WARNED. Now the crew of the Argentine spy ship *Narwal* were extremely worried. Ten days before, they'd been told to stop shadowing the British task force. But they stayed.

We had just finished one op on the West Falklands and were relaxing in the hangar of HMS *Bulwark* when the boss called us forward to the ops room. We were given a quick brief about the situation and all of a sudden it was back to 1 SBS and the maritime counter-terrorist (MCT) tasks. We were going to do a live MCT task in the middle of the South Atlantic.

The Navy frigate captain had been polite but chillingly clear: either the factory fishing ship *Narwal* left the exclusion zone at once, or she risked being blown out of the water. With every day that passed, the men on board must have felt the odds mounting against them. Now, on 9 May 1982, five weeks after the massive amphibious Argentine invasion of the Falklands and with the British forces gathering for the counter-punch, the *Narwal*'s crew knew they were living on borrowed time.

After the warning, the spy ship's captain repeatedly requested permission for his ship to return home to the Argentine. Each time, the Argentine Command in Puerto Belgrano came back with the same reply: the factory fishing ship *Narwal* was gathering vital intelligence. One thing they'd learned is that the Royal Navy ships spent a lot of time with their radars shut down. Total EMCON (emission control) is necessary because an active radar set can be detected at a longer range than it will pick up the enemy. If you have more than one ship (or plane) monitoring an area, you can compare the bearings to give a pretty accurate location of where the radar set is transmitting from. The British ships had their 'passive' systems running of course, so they would pick up any radar signals from the Argentines. It all turns into a gigantic game of blind man's bluff and, given the lethality of modern anti-ship missiles, whoever gets spotted first usually loses.

The Argentines had tried various ways to get a fix on the task force. They'd sent civilian Boeing 707s far out into the Atlantic, using their standard civvie search radar to see what they could find. Sea Harriers warned at least one of these away en route to the Falklands and they ceased hazarding these big jets once the shooting war began. They tried once more with a civvie Lear Jet but a Royal Navy destroyer shot it down at 40,000 feet with a Sea Dart missile. No survivors.

So the *Narwal* had to stay. It was just too useful to the Argentine high command for them to abandon the mission. They assumed (correctly) that the Royal Navy would not sink an unarmed trawler, even if that trawler was spying on them. The British, after all, had invented the concept of fair play. British reaction was cautious at first; after all, the Navy was used to its exercises being shadowed by Soviet spy trawlers, so it had plenty of experience of disguising what it was up to, even when the opposition was watching. But now it was time to take action.

The day was typical of the South Atlantic for that time of year: overcast, with a ferocious, bone-chilling wind from the south-west, and sharp spatters of freezing rain lashing at the long steep swells of the running sea. Now, with the light fading, the spy ship's watch on deck was anxiously scanning the sky through their binoculars. As always, at dawn and dusk, they dreaded the sight of what the Argentine pilots called 'the widow-maker': the Sea Harrier.

We made ourselves ready. The four Sea King Mark 4 helicopters were burning and turning on *Bulwark*'s deck, ready for the teams to board the 1,300-ton fishing vessel. From all accounts the Duke of York was flying one of the Lynx helicopters. Was this true? I would like to think so, as we had heard many a story during our short stay on board that all the duke was used for was in the helicopter off the port bow monitoring for enemy submarines. In total a sixteen-man team deployed in the four helicopters, split into one assault group of eight men which occupied two helicopters and two helicopters (to be used as gun ships) which contained four men each. Within the gun ships we made use of the big open doors and placed the GPMGs that would offer covering fire when the assault teams started their assault. All the helicopters were fitted with 'fast ropes' just in case the assault teams were unable to make it – a form of back-up. The engines thundered into life, the vibration increased and we were airborne and moving towards the spy ship.

They still didn't see the first tiny speck coming at them low and fast from the junction of the sky and sea. Nor the second one that joined it a few breaths later. In fact, it wasn't until the lead aircraft popped up to 500 feet, slowing to 400 knots for its attack run, that they let out their first shouts of alarm. By which time, for the *Narwal*, it was far, far too late.

The lead Harrier's pilot put the fighter-bomber's nose down the optimal 20 degrees for his strafing run. At just under half a mile he opened fire, walking the white froth of his shells from the sea and on to his target. The fat 30 mm projectiles from the twin Mark 4 Aden cannon-pods under the aircraft's belly ripped in through the *Narwal*'s metal sides, wrecking her upper decks. Moments later, the second Harrier made its follow-up attack, holing the hull, and the spy ship stopped dead in the water. It started to roll and yaw in the rough, rolling seas that slowly got worse.

Now it was our turn.

We were to execute a fast-rope helicopter assault, and our mission was simple: to secure, and in the case of armed resistance, kill the spy ship's crew. The British intended to land at San Carlos water soon and the *Narwal*'s intelligence was too good by half. Our job was to stop her – permanently.

We watched the Sea Harriers attack from the helicopters. For the assault itself we were using 60-foot lengths of rope attached to dedicated strong points on the Sea Kings, each rope with a breaking strain of 750 lbs. My team was exiting from the starboard side door of the helicopter, assaulting the target's stern. The difficult thing would be to get that plaited Terylene rope on to the deck, first time, while the helicopter maintained exact station over the ship. Against us were the wind and waves – for us was some extremely good intelligence work.

The Sea Harriers had made photographic passes over the *Narwal* many times before. Not only did we have candid camera close-ups of her decks, we had also obtained the plans of the ship down to the last inch. We'd chalked out a scale drawing of the Argentine ship on the hangar deck of the *Hermes*, rehearsing our assault on this outline until we had it down pat. We were determined not to foul this one up. We'd practised this type of attack a hundred times, more, on friendly shipping on the high seas. But this was war. And it was the first time any of us had done it for real.

Fast-roping techniques are simple in theory: we use a two-handed grip, one hand reversed on the rope as a brake, the other gripping normally. We wear special asbestos-padded gloves to withstand the friction burn, and we go straight down, no messing. It takes a certain amount of upper-body strength. On the way down, you push the rope away from your body, get the legs shoulder-width apart, and look down between them so that you can see at all times what is happening. No feet on the rope at any time. That's why it's fast.

The Harriers had done their bit. I heard the call come through in my earpiece: 'Go! Go! Go!'

We came in from downwind at 40 feet, the Sea King making a near-maximum 120 knots, and I could see it was rough, the waves leaping up to meet us, long cold tendrils of white spray riffling off their crests. I felt a lurch of anxiety looking down at that: a near-gale would do nothing to increase our chances of getting the rope on that deck on the first pass. This was a small target, and the smaller the ship, the more difficult to get aboard hard and fast. As usual, when I was nervous, I felt a strong desire to laugh.

'Make ready!' I shouted.

We cocked and released the bolts on our MP5 sub-machine guns.

'On safety!' Never shoot down a rope with the safety released.

Along with the HKs we had flash-bangs, or thunder flashes, 9 mm Sig-Sauer pistols, and CLC: linear cutting charges for blowing open any barricaded steel doors or hatches.

One minute from target. The pilot eased back on his collective stick as we approached the target. The Sea King put its nose up as it arrived over the ship. The door-loader kicked out the fat coil of green rope. Engines, rotor-blades, the sea, the door-loader – all were roaring in my ears as my hands firmly gripped the top of the rope. Through the opening, I could see the door-gunner on board one of the support helicopters crouched over his GPMG, ready to suppress any incoming fire. Then the covering fire started. Machine guns clattered away, splashes rose around the hull and I could see bits fly off the deck and superstruc-ture. I checked my hands holding on to the rope again, hands on, ready to go. The spy ship pitched and tossed beneath my feet.

The *Narwal* hove to, rolling heavily in the swell. She looked robust and seaworthy, but she was beam on to the sea, which suggested she was out of control. No captain in his right mind would put up with that sickening motion. What with the wind whipping across the deck and that wild roll, the worst happened, exactly as I'd feared: the rope missed on the first drop.

Fuck! I thought. Too slow! Get it on!

The door-loader leant out on his stomach, screaming instructions to the pilot as he pulled the rope back into the helicopter. No sooner had the rope been recovered, it was out again. More covering fire went down as the teams appreciated the situation. The rounds were slamming in the superstructure of the boat; doors were swaying to and fro as the 7.62 bullets pounded the open openings.

Whump! The heavy coil hit the spy ship's deck at the second attempt. We were out and down.

There was no one in sight. I flicked off the MP5's safety-catch and waited for back-up. When Joe arrived we ran for the bridge. It was small, about fifteen feet wide by ten in length, and when I reached the door I stopped dead. The Harriers had mangled it. The steel bulkheads were peppered with 30 mm cannon holes, the windows and the interior smashed to pieces. But it was the blood that halted me in my tracks. The entire deck of the bridge was awash with blood, mingled with seawater and broken glass. The red watery mess backed and swilled as the *Narwal* barrelled under my feet.

Then I saw the captain. He was lying in the corner, with his eyes wide open. Dead. Both his legs were shot away. He'd been blown off his chair by the cannon shells and bled to death where he'd landed. So much for the advice from Puerto Belgrano.

The helmsman and the radio operator cowered away from me, their hands in front of their faces, shouting something in Spanish that I didn't need to be a linguist to understand. These two were also losing blood, but they'd been lucky: they only had superficial shrapnel wounds. The rest of my group was up with me at my shoulder. Glancing back, I could see the last man of our second assault team fast-roping to the deck.

I left two men to secure the bridge and give first aid, the rest of us splitting into two search teams as planned, to make an assault clearance of the ship fore and aft. Some of the crew were hiding in their bunks and in cupboards. Others were just standing in the galley, making no attempt to resist. We went through the trawler fast, no time for the niceties, flushing them out and ordering them up top, shoving the plasticuffs on them. We had zero time to waste, because of the fuel endurance restrictions on the helicopters. And there was another reason for extreme speed: the *Narwal* was beginning to list. I could feel her settling under my feet: she was holed beneath the water-line. Time to get off. The ship was cleared and we took custody of the charts and the ship's log. Her log clearly showed she had been ordered to shadow the fleet and pass that information to Argentine intelligence.

We had the whole remaining crew of eleven men to get up into the reception Sea King, including our star prize, the Argentine Naval Intelligence officer we found aboard. He'd been collating and sending all the data. We winched them up one after another. They were cold, utterly demoralised, and very frightened. It took an eternity to get them all up. The sea state was getting worse all the time as the wind increased. I cursed under my breath, helping to whack the strops over their heads as each man came up for the lift while the remainder of the team covered the unexpected. At last, we were all back up in the air, heading for the *Hermes*, where the POWs would be handed over for processing.

I took stock of our enemy on the way back. Sitting there, handcuffed, in shock, surrounded by black-faced SBS men armed to the teeth, having just had their ship shot from under their feet and been assaulted from the air, each and every one of them had shit and pissed himself. We had to sit there in the stink all the way back. Then, when we reached the commando carrier *Hermes*, the prisoners refused to get off. They were too afraid of what we'd do to them. In the end, it came down to the threat of our rifle butts to make them budge and we had to drag them off. They were wet and cold, they smelled of urine and shit, they were a long way from home and they crouched down like wet puppies. I felt sorry for them and, indirectly, their families as the Marines dragged them screaming into an isolation area where they were stripped, searched and, under the Geneva convention, looked after.

The Harriers came back and sank the badly listing *Narwal* the next day. She made good target practice. We took the 'steering wheel' of the *Narwal* and to this day it's hanging in the Frog Inn at Poole.

We met the *Narwal* crew some days later; they were different people, and extremely happy; shook my hand, hugged us and in their broken English thanked us. The next day the fleet moved into San Carlos water – the final push had begun.

14 The Spies that Took the Surrender at Pebble Island

W E'D BEEN IN THE FIELD for three weeks. We were unshaven, our hair was long and matted, and we smelled dank. It was impossible to tell, given the filthy and ragged state of our uniforms, which army we were in. We were also utterly knackered. We clambered into a Sea King that was to take us to our new forward operating base, HMS *Hermes*. We didn't know, when we piled in with all our gear, that this was Rear-Admiral 'Sandy' Woodward's personal flight. He didn't like the look of us. Who were these rough-looking smelly types carrying enemy weapons? What were they doing on my helicopter?

The admiral asked his staff. They couldn't give him any answers. This wasn't surprising: for Opsec reasons, details of our missions were on a strict 'need to know' basis. We sat there, giving it the 'thousand yard stare' because we'd had next to no sleep in days. They interpreted this as the glaring face of enemy commandos. For 150 miles we flew, little knowing the consternation we'd set in train.

The first indication that something was wrong came when we put down on the *Hermes*. The two Royal Navy sonar crewmen in the rear cabin disappeared. Then the door-loader asked us to place our captured Argentine SLRs and an AK-47, which we were taking to the *Hermes* as souvenirs for the crew, at the back of the helicopter. But he didn't ask about our pistols. Odd, that.

I got up to disembark, but the doors didn't open. Once Admiral Woodward and his band of merry men disembarked we seemed to sit there for hours. Suddenly, the sliding door was flung back, and we found ourselves confronted by at least twenty armed Marines. They had their Armalites pointed straight into our faces. We looked at them, and I thought this was a great wind-up. Great joke. The team and I even started laughing. But we soon stopped when we saw they were serious. Deadly serious.

They ordered us off the Sea King, dragged us across the flight-deck, and spreadeagled us against the nearest bit of superstructure. They were arresting us! They searched us and found our concealed pistols – not standard issue British ones, either. Then they realised we carried no ID, no dog-tags, and that confirmed their worst suspicions. Every time one of us opened our mouths to protest, we were ordered to shut up and a butt of an Armalite was stuck between our shoulder blades.

We lay there for an hour, while the command tried to make a decision about what to do with us. It was only because an officer I'd taken through his SC3s course (Selection) recognised me underneath the encrusted grime on my face that we were able at last to tell them we were Brits.

'Don, is that you?' he asked.

'Yes, George it is. Can you tell these wankers who we are before these naval officers do something stupid!' I had to talk fast before they clobbered me with a gun again.

'Sure, mate, wait here.'

'Well, I can hardly go anywhere else with a fucking M16 pointing at my head, George!' I shouted. He chuckled and ran to the operations room. At last we were released.

This was an example of wild, paranoid overreaction, and it all came from Admiral Woodward's initial response. It's fair to say that he should have been briefed by the loader that we'd be on his flight. (I sometimes think it was a point being made by Woodward who really was not happy about us being on his personal flight – typical attitude of a senior naval officer.) But even so, common sense should have prevailed. It was just not possible that a party of eight Argentine SF men could or would get so far down the line without being detected, and then, having waltzed on board his flight, fail to kill him. The whole episode was ludicrous. Sad, but true. On completion of this bizarre scenario, it all suddenly turned into laughs and jokes. We had brought the Argentine weapons on board for a reason: the relevant messes had asked us nearly three months ago to bring back some memorabilia. And we did because they looked after us so well prior and during our deployment.

Another example of a paranoid attitude was the attack on Pebble Island by the SAS, using SBS techniques. An operation that never got to the SBS command because the SAS had bulldozed their way into the Falklands War. Fortunately, it went well.

It all started when some of the Sea Harriers returned from a mission to Port Stanley. They had detected radar emissions from the Pebble Island area. D Squadron SAS were tasked to mount a recce. So, while we were taking out Fanning Head, the SAS boat troop sent two Kleppers with their two-man crews to East Falkland, and then paddled across to Pebble Island. The current made this a tricky operation, but they did well, got there undetected and cached the boats. They set up an OP and reported what they saw. It was indeed an improvised air strip, with Pucara ground attack aircraft lined up ready to roll, plus a Shorts Skyvan and five Aeromacchi jet trainers. Some of the planes were actually mock-ups to confuse aerial recce, but the others were real enough.

An operational Argentine air base there could really spoil our chances, so it wasn't surprising that the entire D Squadron were promptly warned off for an attack. The recce party was told to man the RV and that the squadron intended to come ashore and destroy the airfield radar, garrison and aircraft.

The SAS prepared explosive charges that would be used to destroy the aircraft. Plenty of ammunition was assembled, with each man carrying a 66 mm LAW (light anti-tank weapon) and a pair of 81 mm mortar bombs in addition to his own kit. GPMGs, M16s and SLRs were carried and MP5SDs – silenced sub-machine guns – were brought to take out enemy sentries. The support group of mortars had about 80 rounds of ammo for immediate fire support, but the real firepower would come from a frigate offshore.

The plan called for 18 Troop to destroy the airfield and aircraft; 19 Troop was the reserve demolition troop; 16 Troop would take out the garrison; and 17 Troop would be split to assist the others. A separate party would direct the NGS and the recce group would act as guides.

At this time the fleet was outside the TLZ (total exclusion zone) and would have to steam hard towards the Falklands to enable the helicopters to fly them to the island. The TLZ was 200 miles and the fleet would only go half way. They would fly the rest. The weather was not with them: it was stormy with a heavy swell. At the designated time that evening the squadron were ready to depart, but unfortunately the Navy was not; they had not made progress as fast as expected. The helicopter could not take off with the wind so strong across deck, and the ship would not slow down for a further hour. The squadron eventually departed in two Sea King helicopters, each carrying about twenty men. The flight took 90 minutes, the pilots flying with NVGs at low level.

Five miles out we got an update from the guys on the ground in the OP. There were still eleven enemy aircraft there. The garrison appeared to be asleep at the switch: no sign of enemy patrols. Down we went and were guided to the FRV (final rendezvous). We knew we were horribly behind schedule now. We'd have an hour at most on target before it was time to escape back into the night. For good reason, the Navy did not want to be caught offshore by daylight. The Argentine Super Etendards were waiting with their Exocets.

No. 18 Troop had still not reached the FRV but we couldn't wait. So 19 Troop took their task of attacking the airfield and aircraft; 16 Troop still had to deal with the garrison. This caused a bit more time to be lost, so we split up and pushed on to do our tasks. The mortar was set up and 19 Troop moved around to the airfield. In 16 Troop we reckoned that the majority of the Argentine garrison would be in the sheep-shearing shed, but we intended to approach as close as possible; we had MP5SDs for any sentries; we also wanted to knock on the nearest house and ask the occupants where the enemy were billeted. (After the conflict it was confirmed that they were in the sheep shed that night but moved out afterwards.)

The attack was started by 19 Troop. They had a great time, running around in pairs torching the aircraft by whatever means they could. The purpose-made explosive charges were still with 18 Troop, wherever they were, so 19 Troop took out the planes with hand-grenades, 66 mm LAWs and machine gun fire.

There was no immediate reaction from the Argentines. No. 16 Troop was still near the sheep-shearing shed when we were told to return to the FRV. Ten minutes later they were told to carry on with the mission. Then, 400 yards away from our target, they were called back again because 19 Troop had finished their task and were pulling off the airfield. Make your minds up, guys! Then the enemy woke up and set off some demolition charges that cratered the runway and injured two members of 19 Troop. There was a brief firefight as the injured personnel were recovered, but the Argentine fire slackened off. Once they knew the airfield was clear of Brits, the mortar troop let fly to keep the enemy's heads down as the raiding party scrambled back to the FRV. Naval gunfire support, in the shape of a frigate's 4.5-inch gun, was directed on to the airfield and surrounding hills. Soon, everybody was at the FRV, including 18 Troop. The two helicopters returned to pick up the squadron and returned them to HMS *Hermes* as it was getting light. The Klepper cache was picked up after at the end of the campaign.

After our short but exciting R&R on *Hermes*, we were sent back into the field. Back to Pebble Island in fact. My team was briefed to conduct observation of the airfield, prior to the possible surrender of the enemy garrison. We were moved in that night by a Sea King helicopter about four miles north of the airfield. We had been there for about seven days when we were told that the Argentine forces had surrendered. 'Take the surrender at Pebble Island', were my orders. There would be another team landing sometime in the morning, who and exactly when was anyone's guess with the war over and the celebrations already under way.

The worry going through my mind was: Did they know they were supposed to be surrendering?

By this stage, I had zero confidence in the Argentine chain of command. There were four of us in the OP 700 yards from the airfield. There were more than 130 Argentine troops, armed to the teeth, and dug in with tripod-mounted .30 machine guns. We were told the Argentine commander of Pebble Island base was aware of the surrender. We had to believe it.

The trenches they'd quarried all around the area had been abandoned, which had to be a good sign. There were about fifteen Pucara aircraft dotted about on the field, some of them damaged beyond repair, the rest with their instrument panels smashed and their compasses ripped out by the Argentines, deliberately, to leave as little for us as possible.

We looked south-east, along the runway. The civilian settlement was at the far end of that. We moved in to take the surrender at 0900 the next morning, keeping strictly to the main track up to the gate in the perimeter fence. There was a danger of mines, liberally sown by the enemy all over the Falklands and not marked on any maps. We moved in closer through the gate and past the perimeter fence that surrounded the small settlement. It was quiet and there did not seem that many people around. Where were all those soldiers we had seen for the last week? The weather was bad, but not that bad for a hundred soldiers to hide away or disappear. I was sure we had stayed awake the previous night, and we certainly did not hear any air movement during the previous day.

The door slowly opened and a number of people came out of the biggest building within the settlement. Just a handful of blokes at first. They started walking towards us. Then others followed. Every door we could see opened, and out they came, hordes of them. Thankfully, I had left Steve and Terry with the GPMG up on the high ground to our rear and right, just in case something went wrong.

So here we were. Face to face with enemy at last. The nearest man wore the insignia of a colonel, the man to his right had on what looked like the badge of a sergeant major or RSM. Neither of us spoke a word of Spanish, but the colonel turned out to be able to speak passable English.

'Good morning, Colonel,' I said. 'My name is Sergeant Camsell, and I'm here to take your surrender of this island.'

He was courtesy personified. He was about my height, only much slighter in build, in uniform but without his cap. He seemed relaxed, even relieved. But what added to my nerves was the way his entire command, all 130 of them, were lined up behind him as we spoke, all heavily armed.

'I'd like all your men to . . .' I began, but at that moment our reinforcements, all four of them, finally arrived on the Lynx. It was the RSM of the service, just to add a bit more punch to the surrender. The colonel turned and fired off a command in Spanish. Slowly, and with great reluctance, his men filed past placing their weapons and ammunition on a pile that was soon as high as a house. Once that was done, we lined them up, counted them, and took their names. It was then, after a long wait, they were taken to Stanley for evacuation by Sea King shuttle; it took up the remainder of the day. Within that day, that seemed to last for ever, we became quite friendly with the more senior officers and men. War is a sad thing as there is always a loser. Thankfully, Britain has never been in that situation.

The islanders hugged and kissed us and we experienced once again the happier side of war. One particular point sticks in my throat. Before the helicopters arrived to evacuate the Argentine soldiers, a fresh-faced lady of about twenty, with long blonde hair, ran up and hugged me. Tears streamed down her face. She couldn't stop saying 'thank you'. She took my hand and gave me a handful of pebbles: all of mixed colours and shapes, shining in the afternoon sun.

'These are for you,' she blurted in between the tears. 'Please keep them and remember this day as we will never forget it.' Then she ran back to her house and waved us goodbye. The lump in the throat was there for me: a moment I won't forget.

Once we'd loaded the captured weapons into cargo nets, the helicopters came in and whisked them up, took them out to sea, and dumped them, except for specific and more useful weapons, which were allocated to various units.

We searched carefully, as from all accounts the outer perimeter of the settlement was covered in mines. The RSM was very strict in our movements; as he stated, the war was over and we must maintain the attitude that we had maintained throughout the campaign. After all we did not want any injuries now so close to the finishing line. We found their ammunition still packed and sealed in its boxes. Large stocks of new night viewing goggles (NVGs), unused because, from all accounts, they did not know how to use them. Again, we had the impression they thought that we'd never come to fight them.

We went on the last Sea King to Stanley, the first time we'd seen it. It reminded me of an ant hill, with all these British and Argentine troops swarming all over it; troops in every street; military police taking charge of the mass of vehicles and men; huge piles of weapons strewn around the airport; craters, evidence of the Vulcan bomber raids.

We started our long trip home firstly to the Ascension Islands on the Royal Fleet auxiliary *Fort Grange*. We were delayed sailing for many days due to the situation in Stanley. It was one long, fantastic party, with the whole squadron on board. The SAS turned up alongside on another ship. We invited them over for a drink. Initially we got on extremely well until some young and pissed badged regiment man mentioned the 'blue on blue' and the killing of our great friend Kiwi. (An SAS patrol had shot at an SBS patrol that strayed into their area, killing one of our men.) It was not something to joke about. The regiment did not come back again.

We were not allowed a flowery return to UK, in fact even the wives were told that they were not to come and meet their husbands. Of course the wives disagreed and promptly hired a coach and drove to Brize Norton, forced their way into the airbase and were there to meet their husbands. We were offloaded from the VC10 and held up at the customs hall where a customs officer greeted us with these exact words: 'Gentlemen, firstly congratulations on the fantastic job that you achieved in the Falklands. Secondly, I could not give a flying fuck how much drink or cigarettes that you bring in. However, gentlemen, weapons and ancillaries are a no-no; please be sensible and declare them with no backlash to you at all.'

What a gent. (We smuggled in the weapons anyway as the Brize Norton armoury was no doubt full up.)

We walked out and the wives were there to meet us. Fiona was with her dad, ex-Parachute Regiment and Arnhem veteran: he knew how we felt. When we got back home, I was overwhelmed to see that the entire street where I live in Poole had put out bunting and flags, and organised a street party in my honour. When I saw this, with Fiona and the children there, I choked up. It was extremely embarrassing, extremely good, and I still get all hot and bothered when I think about it. It was only when I got home and thought about the things I'd done, and got away with, and remembered the guys on the *Fearless* dying with their limbs blown off, that I knew how lucky we had been.

15 Drug Wars

OR MOST OF MY TIME IN THE SBS we were preparing for
World War III. When the Soviet forces came pouring
westwards, we'd be ready. Counter-terrorism and
other missions were valuable training, but our primary
role was to win time for NATO's conventional forces to
deploy into Europe to stop the Red Hordes. With the
collapse of the Evil Empire, were we out of a job?

The end of the Cold War has not produced the peace
dividend widely touted in the early 1990s. The new world
order is, if anything, more dangerous than the days when
NATO and Warsaw Pact armies confronted each other across
the inner German border. Without the restraint imposed by
the rival superpowers, wars have flared across Africa, Asia, the
Middle East and the Balkans. Public pressure for intervention
has led to a number of 'peace-keeping' missions where peace
has to be imposed by the threat of superior firepower. These
situations have created many possible taskings for the Special
Forces. In addition to these fairly conventional missions, UK
Special Forces have joined the war on drugs. Just as MI5 has
diversified into other areas of civilian intelligence, the SBS
approaches the end of the century with one eye on the future
– ready to work with other government agencies.

The National Crime Intelligence Service (NCIS) may
well become one of the organisations seeking SBS support.

Launched in 1992, it targets the higher echelons of crime. With about 500 staff, drawn from the police, Customs and Excise, and the Home Office, it was one of the first services to be set up in Europe to provide criminal intelligence on a national scale. Its international division manages a network of European drugs liaison officers and is linked up with the worldwide DLO group managed by Customs and Excise. The UK Bureau of Interpol is also based within this division, providing NCIS with direct access to Interpol's 176 member countries.

Although all projects in which the SBS are involved are security sensitive and remain concealed from public view, modern tasking is leading them towards longer-term involvement in what may be regarded as ultra-secret activity. The SBS is never glimpsed by the media and especially not by cameramen, although one of its earliest 'civilian' tasks in the drugs arena did make the headlines, simply because of the size of the target: a ton of pure cocaine worth £160 million, the largest quantity ever seized in Britain.

Months of monitoring the movements of a ship, 'Fox Trot Five' and its mainly British crew, culminated with a spectacular raid at Greenwich, London, with an SBS team swarming all over the vessel as she tied up on the Thames. The boat, which had been bought in America, had sailed to an island off Colombia and was tracked across the Atlantic. On 23 November 1992, she was sailing towards a mooring beside a warehouse on the edge of the Thames.

The vessel sailed on along the south coast and back up into the Thames, where she was once again moored at Greenwich. There, in an operation which included Customs and Excise, Interpol, the US Drugs Enforcement Administration, Scotland Yard – and now the SBS – the trap was sprung. The crew scuttled in and out of the ship lugging fat bundles, wrapped in black bin liners. Inside the warehouse, the heap grew and grew, watched from a discreet distance by a number of observation posts.

Suddenly, two RIBS carrying SBS teams roared across the Thames from the opposite bank. In complete counter-terrorist kit, the assault teams raced up caving ladders and, within a few seconds of the signal being given, the ship's deck was alive with 'men in black'. The crew didn't know what hit them. One moment they were quietly unloading their cargo, the next they were staring down the wrong end of an HKMP5.

The SBS teams were followed by a large contingent of armed police and customs men. The police used a JCB to smash down the doors to the warehouse where they arrested five men. Within a matter of hours, some 200 policemen raided 18 different addresses in the south-east of England. The entire drug network found itself in jail.

The SBS faded quietly into the background. But for a sharp-eyed woman named Joyce Lowman, who took a photo of the SBS team as it boarded the boat, our presence would not have been revealed.

The drug barons have come to know and hate us. The latest drug dealer to have his evil business shattered is someone who started out in Middlesbrough, but muscled his way into the drug racket until he was rated one of the top 25 narcotics dealers in Europe. He was a walking cliché: £15,000 Rolex on his arm, crocodiles in the swimming pool at his Spanish villa . . . obviously, image problems beyond the dreams of analysts.

His drugs were coming over in an ocean-going tug, 90 feet in length and broad in the beam. Originating in Venezuela, the ship had plotted an irregular path across the Atlantic, making for the west coast of Scotland. We had an A1 source she was drug-running, and the decision was taken to hit her at sea the minute she got inside British territorial waters. For us, it was a case of business as usual. We'd rehearsed a number of similar operations that had all been cancelled at the last moment. This time, we went to a holding area at an RAF base in Scotland, one we've now become very familiar with.

Customs and Excise had overall control of the op, but we were the cutting edge. We had a couple of customs officers with us to make the arrests; we'd trained them in fast roping and some other counter-terrorist skills. I led the eight-man assault team. The hangar was a hive of activity as we prepared for our first big hit on the drug scene. We missed the usual small group from Hereford, which was no bad thing. To our surprise, they had not forced their way into this operation. As these operations progressed, we realised why. If there was a particular job of this nature only the participants involved were briefed, and in later operations personnel were not briefed until they were out of range of mobile phones far out to sea. As we were to find out, drug barons might not have the military firepower of the Warsaw Pact, but their financial firepower can have a corrosive effect on any organisation that takes them on. 'Virtue is insufficiency of temptation' according to Thackeray: and guys able to ship over £100 million worth of dope could offer temptation beyond dreams of avarice.

We trained the customs team and rehearsed the complete operation again and again with the same Chinook crew. At last we got the signal. The RAF Nimrod shadowing the ship reported that she was approaching British waters. Final briefing complete, the CH 47 lifted off and headed on a course that would intercept the trawler from the rear. The flight took about forty minutes, the helicopter buffeted by the strong wind, rain lashing against the fuselage. Just the way we like it.

Visibility was poor, no more than 1,800 yards. We spotted the tug, a tiny thing on a big ocean. I felt the adrenalin surge through me; it wasn't the sudden descent that put my stomach in my mouth. No one knew how these guys would react. We were going on past history and were prepared for a gun battle. Unfortunately, I prefer gun battles to be conducted on the ground, not when we are flying down a 40-foot 'fast-rope' during a force 6 gale on to a bobbing cork in the Irish Sea.

The CH 47 moved into the hover above the tug. The fast rope shot out and, for once, it landed in the correct place. We had to get guys on the deck soonest. The first man whizzed down the rope, then the second. No gunfire. I was happy: we had a good foothold on the ground, all the team on the target and still no aggressive response from the so-called smugglers. The team moved to the bridge and accommodation; the CH 47 moved to the rear of the tug, just out of small arms range, but able to maintain radio contact with us. The crew just stood about with gaping mouths. They thought they were safe out here: the risky bit would not come until they tried to land. They were all arrested and the search began.

The ship was crammed to the gunwales with cannabis. Not a single inch of space had been left empty. We could hardly get inside the doors. She had a crew of eleven, mostly Lascars, several of whom had no ID papers and spoke no English. The tug was steered into the nearest port where a larger customs team waited. We were delighted it went so smoothly.

Our next anti-drugs mission took place off the Isle of Wight, by boat and helicopter. Hampshire police led this particular operation and again the security needed to be tight, although it is difficult to be secure in such a place as the Isle of Wight. It's a small community in which everybody knows everybody else on the island. Outside the tourist season, there aren't many visitors. In the dead of winter, the presence of the SBS, not to mention customs, police and other officials would be pretty hard to conceal. And we had no idea who might be in the pay of the drug runners. Apparently, the smugglers had got wind of a number of customs operations in the past. For this one, the police had to bring a lot of kit to the island on the ferry: they admitted this was far from ideal, but they didn't have any other option.

The Isle of Wight was a useful entry to the UK for the drug runners: lots of little bays and beaches, close to the shipping lanes and pretty quiet off-season. The dope would be landed by small boat and, when the smugglers thought it safe, they'd bring it to the UK on the ferry.

The RIBs were in position. A Lynx helicopter had picked up the suspect boat twenty miles from the island. The RIBs waited in ambush as the Lynx's radar tracked the boat across the English Channel. Meanwhile other teams ashore covered every possible landfall and approach. If the smugglers made it to a beach, they'd meet an interesting reception. The problem was that the teams had to move openly to get to the likely beaches in time. That meant a lot of vehicle movement, unusual for the winter, and obvious to a local who might have been watching.

Whether they were tipped off by an observer ashore, or just kept a good look-out, we never learned. The Lynx reported that the smugglers had changed course and were on their run in to a small inlet that looked ideal from their point of view. The RIBs were directed by the helicopter to attack in the usual way, but this time they saw us. One moment the speedboat was pottering along, headed for the beach, then suddenly two great balls of foam surged in her wake. They flung the throttles full open and raced for the open sea. The RIBs did the same, their powerful twin engines flat out. The hulls crashed through the waves with a sickening thump, thump, thump, everyone clinging on for dear life. You try to take the shock on your knees for as long as you can, but it jars right up your spine whatever you do.

The speedboat skidded around the inlet, her massive bow wave visible to everyone ashore. Only a real racing speedboat can outrun an RIB in experienced hands, but a stern chase is always a long chase. It didn't matter that we saw them throw stuff overboard: by the time we overhauled the speedboat and forced it to heave to, the drugs were gone. They'd taken the precaution of shipping them in weighted bags. The evidence was on the bottom of the English Channel. So the lobsters got stoned and the police couldn't make any major charges stick on the guys we arrested.

The rush of storming a boat on the high seas has to be experienced to be believed. But every now and then, you get a savage reminder that this is a very dangerous business. I suppose it was inevitable that someone would get hurt or killed on one of these missions, but when this one began, on a Friday afternoon as usual (these people loved working on the weekend – it totally fucked ours up) we had no thought beyond getting the job done.

Two detectives pitched up at Poole in civvies and were ushered into the boss's office. After a long session behind a locked door, I was invited in to be briefed on the operation. It was drug related, I think our fourth such operation that month. The target ship was coming in from the north, bound for the west coast of Scotland. The police needed our expertise not only on the water but on land.

I briefed the team and off we went to our usual hangar on an RAF base. Blue vans and RIBs slipped into the base quietly, late at night. The air support arrived in the early hours of the following morning – the team was complete again. Our Nimrod 'eye in the sky' monitored every movement of the ship. For ten days, Nimrods on station over the North Atlantic kept it under radar surveillance during their eight-hour maritime patrol flights. Twenty-four hours a day, there's a Nimrod up there, scanning everything on the surface, in every direction, on its Searchwater radar.

We waited. And waited. We rehearsed the thing to death, and rehearsed some more to pass the time. Nothing definite was coming out of the head shed and it seemed that nobody could make a decision. The ship was skirting British territorial waters, but seemed reluctant to take the plunge. Did they know we were on to them? Or were they just being cautious?

The game of cat and mouse continued until the accountants called us off. What with the Nimrod patrols, police overtime and a hangar full of commandos, there was a lot of money getting spent on this operation and no sign of it coming to a conclusion. The ship kept in international waters where we were powerless to do anything.

Once the ship turned to head back across the Atlantic, the authorities gave up and we were sent home. We were just glad someone had finally made a decision. No more hanging around. The blue vans headed south for home.

No sooner had we were settled in Poole and our beds, when the target ship entered British territorial waters. Rather than rush us back up there, the decision was made that the Customs and Excise men would try and board the ship from their own boats. The seas were moderate and the future weather forecast was good, so it should not have been that difficult. The team approached the target vessel, but were compromised.

The drug runners reacted smartly. They set fire to their own ship with pre-positioned flare boxes. Obviously, not all of the crew knew of this plan as some panicked and one man leaped overboard between the customs boat and his own ship. A customs man dived in to rescue him but got crushed to death when the two boats suddenly crashed together. The target ship was eventually overhauled, fires extinguished and crew arrested in possession of vast quantities of drugs.

Apart from the tragic death of one of the customs officers, the case posed another obvious question. Was it merely a coincidence that the moment the SBS team returned to Poole, the drug runners put about and headed for the UK after all? Did we have a traitor in our midst? At the time we wondered about a mole in the various law enforcement agencies. If anyone had suggested it was one of our own, I don't think they would have been believed. Unfortunately.

Some people question the involvement of the Special Forces in what is effectively a police role. It is certainly an area in which the SBS are likely to be used more and more, but they are basically Royal Marines, in other words soldiers, commandos, with specialist maritime skills. We have unique abilities. Clearly, we have been involved in a wide diversity of operational activity during the Cold War. Only during the last decade have we moved into this new environment. Our tasking has become more frequent as international hot spots have become more widespread. At the same time, these missions accord with the popular image of the Special Forces: one of apparent personal glorification and self-satisfaction. This will never change.

16 The Gulf War – Beyond Special Forces – 1990

T HREE WEEKS ACCLIMATISING in the desert at a camp in the UAE. We were scratching around for SF tasks: this was going to come down pretty quickly to a set-piece tank battle, which the coalition forces were going to win. But there was a big push from above to get the SAS and us in on the ground, almost regardless of good operational sense. Some of our officers made themselves unpopular by pointing out this amounted to nothing more nor less than PR.

The Gulf War was a new threat using old solutions. In outline Iraq invaded Kuwait in the August of 1990. Saudi Arabia, United Arab Emirates and other Arab countries combined with NATO countries to form a coalition against Iraqi aggression. The UN and the US voted to support the military action. The military build-up along the Saudi border and the air war started against Iraq on 16 January 1991. Saddam's response was to target the cohesion of the coalition by the threat of the Scud strategic weapon. He was able to operate mobile launchers, which seemed to be operating from the west of Iraq. Saddam decided to target Israel hoping that Israel's reaction would break the coalition; fortunately this did not happen.

The immediate challenge to the coalition was the Iraqi Scuds. From all accounts, Iraq had 30 fixed sites and as many as 20 mobile launchers. However, to find these areas proved an immense challenge. There was 19,000 square miles of western Iraqi desert to be searched and although the coalition had unquestionable air superiority, the AC (air coalition) could not find the Iraqi Scuds. Modern aircraft have targeted guidance systems which are good at finding targets within a mile but not that good at finding targets in thousands of miles.

The solutions that were put in place – for example, patriot air defence missiles – were placed in Israel; in the west of Iraq US SF and UK SAS were launched into long-range desert patrols to locate the Iraqi Scud missile launchers which in some circumstances made things worse. It was stated that only half the missiles would land within two miles of their targets. It was at this time that US special operations and UK SF operations co-located to support the deployed teams. These headquarters were not located near coalition HQ or even the air coalition HQs. This in itself helped the Opsec situation. Only the most senior officers of the coalition HQ had knowledge of any of the missions being conducted. This situation had a down side. There was so much concern for Opsec that it prevented any effective synchronisation of aircraft with the teams.

The date 15 January was the United Nations deadline which ended up being the start of the air war. To follow this, on 18 January the Iraqi launched the first Scuds against coalition targets in Saudi and cities in Israel. The threat from Israel was instant. On 19 January 1991 the Israelis promised to launch their own air attacks and put soldiers on the ground to find the Iraqi Scud launchers. In some sort of compensation deal, the coalition started to insert SAS/SBS road watch teams. But poor staff planning put them into unsuitable operational areas and situations. These teams entered Iraq and with significant special aviation support it reduced the Scud launches to one a day. On 28 February a cease-fire was agreed.

The latter prevented Israel from entering and launching their own anti-Scud operations and the tactical excellence of the operational teams on the ground used stealth in finding their targets and surprise in attacking them. Planning excellence in most cases enabled unit staff to insert operational teams in the right place at the right time, which made a difference for the Army, Fleet and the coalition commander.

I really thought from my personal point of view that the Gulf War had passed us, the Special Forces, by. Although the Special Forces had their own headquarters set-up there seemed to be a lot of in-house fighting once again and every SF unit was desperate for tasks that perhaps were not even there. Once again Hereford were the masters of moving their noses in where it was not wanted. Either pride or stupidity, I am not sure which, made them intent on being seen to do something no matter how many people could be lost – the PR thing all over again. Get men on the plains of Iraq and we will sort the problems out afterwards.

There was one incident as a result of all this glamour-grabbing that still sticks in my throat. This involved an SAS D Squadron commander who was an SBS Officer – he was the first SBS officer to command an SAS squadron under the new joint command procedures; a position full of crevasses, however. They would have to be pretty fucking good to catch this baby out. He was a first-class officer and established a relationship with the Hereford squadron that was second to none. He did have a very good relationship with the CO of the regiment, in fact from all accounts they were very good friends. I personally had some great operational times with him and without sounding a bit sentimental, he proved to be one of the best officers that I have worked with. I am sure others agree.

In outline, this particular officer refused to risk the entire squadron and all its vehicles on a billiard-table desert offering zero cover against insuperable odds. Which is not to say that he didn't try; but his experience on a number of initial sorties into Iraq proved to him that using the squadron in this way was militarily suicidal. But his illustrious commanders kept on insisting 'we've got to get in there, no matter what'. This attitude disregarded any tactical or even strategic sense. It had arisen because of something that has dogged the SAS ever since it shot to prominence during the Iranian Embassy siege: image. And once a Special Forces outfit like the SAS starts living up to its own mythology, it's in trouble.

For refusing to risk the men under his command on an ill-conceived and ill-thought-through public relations exercise, this man, one of the best officers that the squadron ever worked with, was relieved of his command in the field and RTU'd.

It seems that if you don't agree to the command's hare-brained options, no matter how suicidal you think those might be, you can quickly find yourself labelled a coward in the SF. This man was not a coward and didn't know the meaning of the word. He was a first-class officer, who was very shabbily treated and not just by the regiment. I am afraid to say that even our own SBS system apparently had their hands tied to the point that the officer was detached and operating under a separate commander. My immediate thoughts at the time were, What fucking century are we in, the eighteenth or the twentieth? There is a lot of truth in that good old English saying, 'it only happens to the best'.

To prove everyone wrong, he was right. He was replaced by the D Squadron RSM, who duly led them over the border. They ran into the edge – just the edge – of an Iraqi armoured brigade, whose tanks made their Pinky-winky Land-Rovers armed with .50 Brownings and Mark 19 grenade launchers seem just a tad flimsy. The squadron limped back seriously shot up, with three major injuries, and very lucky not to lose any killed. They never went back in that way again and they never reversed the decision either, which was typical of senior command.

Thankfully, you can never keep a good man down; the officer now indirectly overlooks the service, ensuring that none of these harebrained decisions is made in the future.

I was still wondering whether I was right to have committed our people and whether they could achieve anything which our aircraft could not when, on 23 January, Special Forces brought off their first success: blowing up a substantial stretch of the communications network between Baghdad and forward areas. This was a high-risk operation, separate from the SAS deployment, carried out by Special Boat Service (SBS) with great skill, determination and courage in a most hostile environment. The raiding party flew in at night in two Chinooks to a site less than sixty kilometres from Baghdad: in the distance to the north, the sky was lit up by the flames of the capital under bombardment. While men ran in to attack the cables, their helicopters, which had landed a short distance from the main road to Basra, had to keep their engines running – although with rotors disengaged, to cut down the noise – so that they could be sure of taking off swiftly in an emergency. The men then dug down, exposed the cables, removed a length for analysis, placed charges and withdrew, blowing out a considerable section. They also took with them one of the above-ground markers which had designated the cable route and next day they gave it as a souvenir to Norman Schwarzkopf, who was so delighted and impressed by the success of the mission that he immediately reported it to Colin Powell in Washington. Powell in turn passed the good news back to London, so that the first raid made a major contribution towards establishing the reputation and capability not only of our Special Forces, but those of America as well. Just at that moment, in typical

fashion, luck seemed to turn against us; one of the
SAS patrols had been bounced and had scattered.
Storm Command – General Sir Peter de la Billière

The beginning of 'Bravo Two Zero'

We were tasked to go into Iraq and sever the country's key
fibre-optic communications link. There was only one snag:
it was only about 25 miles south of Baghdad.

It was going to be a helicopter insertion, the only way
of getting that close to Baghdad and coming back out alive
again. We'd have to stay in there for perhaps as much as
two hours, locate the fibre-optic link, dig down the twelve
feet of sand it was buried under, and cut it with a shaped
charge. To increase our chances of survival, the two CH
47s would have to remain on the ground with us, burning
and turning, while we did the job. Which had the paradoxi-
cal effect of making the whole operation even more
hazardous, since it is quite difficult to conceal two CH 47s
the size of a small house in the middle of the desert just
outside your enemy's capital. And right next to a major
access road (MAR).

We did our detailed planning and rehearsals at our FOB
nineteen miles outside Riyadh. The helicopter would have
to skirt the big Iraqi armoured formations scattered about
in the area. And it would be slow in the air: there would
be 35 people on the two CH 47s, plus 3 long-range fuel
bladders in the mid-section. Talk about putting all our eggs
in one basket. Because of the fuel problem, the mission
would be strictly time-limited. If we hadn't completed the
task before the cut-off time, we'd have to EXFIL regardless,
failed. And no one in SF likes failing.

The first phase was the insertion. Phase two was securing
the area. Third phase was finding the comms line in the
darkness and then placing the small charge that would blow
the cable apart and cover the mess back-up. The fourth and
final phase was the exfiltration – coming out!

Working for us was the fact that the task was relatively simple, and the fact that the terrain, where we'd be operating, was slightly undulating. There were low ridges and shallow wadis, and we might be able to at least partially conceal the Chinooks. And we'd have top cover in the shape of a squadron of A-10 tank-busting aircraft, standing off ready to jump any interference.

We had six-man protection groups for each element of the operation that was at risk: one group protecting the CH 47s and their crew; one covering the digging team; the rest of us either digging or providing perimeter security.

We left our FOB at about 2030, arriving on the target about two hours later. We flew at ground level, the aircrew wearing NVGs, flinging this great helicopter around the ridge lines in the darkness. It was difficult not to be air-sick. We landed and piled out of the Chinook's tail. The initial security was to protect the big birds as they were our only way out. We cleared the immediate area and then quickly found the tell-tale signs of the cable. So far so good. It was like a well-oiled machine: everything that we did in the rehearsals was being initiated; not one person had to be redirected; they all knew their jobs. We were all sweating, with nerves or just sheer excitement and we had not started digging yet.

I don't know which company laid the cable, but it must have been friendly to our side, because the location we'd been given for it turned out to be exactly right. There was even a small marker post sticking up out of the desert. We were very aware of the road off to our right, which was teeming with traffic, most of it military supply convoys headed south.

The CH 47 left its engine at idle, the rotors turning slowly, so that it needed only thirty seconds to get them up to full speed and take off. In a matter of seconds we were out, deployed and digging. In my six-man cut-off group, which was tasked with taking care of the diggers shovelling hard and fast in the night, we had three GPMGs, a full set of '66' anti-tank rockets, and 203s – enough to discourage all but a full-blown and very determined armoured assault. All the time, we had the distinctive smell of Avgas in our nostrils, and it struck me that this was our greatest concern. We could hear the faint *swish* of the CH 47's rotors as they were moving around. If one of the drivers on that road over there stopped for a piss and smelled the gas, we could be in trouble.

It seemed to be taking them an age to locate the three-inch black plastic-covered cable, even with the aid of the marker post. And it did take more than an hour after several holes were initially dug. But at last we found it, put the small shaped charge on the link, blew it, made sure it was ruptured, and sanitised the area as best we could. We grabbed a small length of the cable to be analysed on our return and also the marker post, which was presented to 'Big Norm'. It was a complete success and one that each individual who participated in it can be rightly proud.

A risky operation, but one that was feasible and worth the risk. Charging out into the barren waste in a Land-Rover with no real objective? That's another matter.

17 Turncoats

A s MORE MARITIME counter-terrorist operations were mounted in support of Customs and Excise, it became obvious that our worst suspicions were true. There was a mole. Information about upcoming drug busts was pissing out all over the place. The details were so timely that the informant had to be right inside SBS HQ at Poole. An undercover Customs and Excise team worked on the case for seven months, trying to find out who was fouling the nest.

The mole turned out to be a Royal Marines Rigid Raider coxswain. He had just left the service, and was not, as reported in the press, a serving member of the SBS; in fact, he'd never been in the SBS. But he did know our methods inside out, having worked alongside us for years. He'd been offered £30,000 to go over to the other side. And he'd accepted this, although in terms of the millions shifted by the drug runners in the UK every year it was a paltry sum.

I freaked when I heard. I knew him, both as a colleague and socially. He'd been a key element in the fight against terrorism and now drugs. He knew everything, from the fuel endurance of a CH 47 to every detail of our weaponry, our kit and our operational procedure. Even worse, as Customs watched and listened, we discovered that he had two other military contacts: another Marine, still serving, and an ex-Para. The pack of them had started working the Spain–Gibraltar cannabis route.

We watched them 24 hours a day for 9 months. Then we found out they had a big shipment coming into the UK. This was personal. We meant to get them.

Certain individuals at Poole were well aware of the situation. Phones were bugged and a number of other suspects were closely scrutinised. From this it was discovered that every move of M Squadron (Maritime Counter Terrorism) was being passed by telephone to an unknown destination.

The man on the inside was a trusted man, due to move to another unit as his draft (posting) at Poole was finished. He diligently fought this proposed move, stating that he was happy where he was. Because he was believed to be a sound and respected operator, the Service fought his case for an extension of his time at Poole. His extension was guaranteed the moment his 'after hours' business was discovered.

We had to mount the operation against their shipment without them finding out. And given the sums of money washing around in the drug business, we couldn't assume that we'd found every turncoat in the organisation. There might be others. They might have powerful friends. Initially, only three individuals knew of the problem. A fourth person was accidentally brought into the equation; there was a team task to the Middle East coming up, and through the team leader's respect for the player, he was adamant that the coxswain was to be in his team going to the Middle East. The team leader was passed from the admin officer to the operations officer and back to the admin officer. The team leader started asking searching questions, so eventually he was brought into the secret. And he would eventually lead the assault on the drug runners' vessel.

It had to be a surprise hit. These guys were doubly dangerous: they knew what they were doing, and they knew the way we operated too. It was going to be tricky to get the assault teams out of Poole without the traitors knowing. The scheme we came up with was to have a team recalled on an individual basis for an exercise in Africa, followed by something in Northern Ireland. This story was widely disseminated and, sure enough, the mole passed on the news to his outside contacts. The team assembled on the forward operating base, in this case a Navy destroyer. At this point, of course, none of the MCT team knew what was going on. Their first hint that something different was planned was when they came aboard to discover their RIBs pre-positioned.

The ship and the team were miles outside territorial waters and all mobile phones of the crew and operational teams were confiscated. No one could make unauthorised contact with the shore base. They were not allowed to have any contact with the outside world during this time, not even a phone call to relatives or friends. For six weeks, the team waited aboard in this state of isolation, while the authorities ashore monitored the progress of the drug runners' shipment from Spain.

We stopped using Royal Marines coxswains. Only badged SBS personnel drove our boats. There was a grim silence during the briefing when we were told that all phone calls on and off camp at Poole had been monitored. But the bugs had done their job. The drugs were on their way while we were believed to be safely away on exercise.

Until that final briefing, the teams had no idea what they would be doing. Only when the 'go' had been received from London were they told of the operation – and that they might know some of the X-rays (terrorists) who were involved in this particular drug swoop. As pictures of the drug group were shown around, members of the MCT team recognised various individuals. Like me, they found it hard to believe we'd been betrayed by fellow Marines.

The dope was aboard an ocean-going tug, its voyage being charted first by military satellite and then by RAF Nimrods, which shadowed it across Biscay.

The destroyer closed the tug during the night, dropping off two 22-foot Rigid Raiders some twelve miles astern of her. Each contained an SBS assault team and two of the volunteers from Customs and Excise that we'd trained. Wherever possible, we had changed our operating procedures from what the opposition might be expecting. Even the timing of the operation. We didn't wait until they were in British waters, but hit them in the middle of the Bay of Biscay.

Biscay was its usual unwelcoming self with steep seas that battered the RIBs as they strove to intercept. Everyone peered through the spray for the first glimpse of the target's masts and superstructure. We were in touch with the Nimrod overhead, getting updates on the range.

'Eleven miles out,' was the word as we slid down one great green mountain and up the next.

'Seven miles.' The wind whipped across the top of the waves, driving salt spray into our faces.

'Five miles – two miles – one mile – where the fuck is it?'

At last we saw a black shape, rising and falling with the gigantic mid-ocean swell; her stern pounded into the black water and rose again, throwing up a wall of water. We were less than 1,600 feet from her. Was anyone watching for us? Were there binos – or assault rifles – trained on us from the small wings either side of the bridge?

We must take it as if there were. Our weapons were at the ready; our pole and ladders hoisted to a height that we thought would be correct. The wind gusted and the rain hit us like a shower of pellets; it was ready to turn to sleet. Perhaps they just didn't notice the echo on their screens, we never knew. But we were ordered to put the poles and ladders away.

Someone in the head shed had the bright idea that we would ride up with the swell – it extended to a good 10–15 feet – and literally step from the RIBs on to the deck of the tug. OK. We watched the waves for a bit, then trusting we'd got the rhythm right, we got ready to pounce.

Up went the Rigid Raider, soaring on the back of a wave the height of a double-decker bus. The moment before it started to subside, we leaped into the air (look, Mum, no safety net!) and landed on the deck. The raider vanished from sight. God knows what would have happened if the opposition had been ready for us, but they weren't. (And we had had to assume the worst: that they'd be ready with flash-bangs, respirators, HKs . . . the whole 'man in black' kit.)

So there we were. It was a few hours before we'd normally mount such an operation, and we were on the deck, weapons at the ready. There was no possibility of back-up if things went wrong. (How many guys were on the ship we had no idea.) But we were armed and very dangerous. We informed the Nimrod that we had 'dry feet', i.e. were on the target ship and not floating in the Atlantic. Then, without a sound, the first two pairs edged across the deck, the snub barrels of their MP5s looking for trouble.

Amazingly, we hadn't seen anybody yet, except for one man on the bridge. They did not hear us coming. We kicked open the doors and charged into the accommodation. The crew had just gone to sleep by the looks of things. One moment they were warm and snug, the next they saw lights and heard shouting. The lights were the powerful torches we have secured under the business end of our MP5s; the shouting was us, telling them to hit the deck. We plasticuffed their arms and legs. The only weapon that was found was a knife. We could hear the stun-grenades going off in the other areas of the ship: nothing like a 'flash bang' to completely disorientate a bad guy lurking the other side of a door. There was no opposition anywhere else on board.

We handed them over to the custom officers who gave each individual that dreaded phrase 'you are under arrest for suspected importation of a class B drug' (cannabis). Their worst nightmare had become reality. Once everything had settled the ship remained on course for the UK under the close watch of the Navy. All in all we had three ex-Marines who worked closely with us in the SBS, and over 26,000 lbs of cannabis. Each bale smelled of fish – they'd been concealed in a Moroccan fishing boat on their way over from Africa.

The two men we most wanted were in their cabins, fast asleep. What was really strange and almost comical was the look on the faces of these ex-Special Forces support men when we turned up in the black kit in the middle of the night and arrested them. Despite all that they knew about our monitoring and assault capabilities, they were utterly surprised when it happened to them. Luckily for them, they were unarmed, except for a knife. Had they been tooled up, they'd have gone down for twenty years plus. As it was, they each got fourteen years.

Our euphoria was short-lived. We'd pulled off a difficult operation, but it stuck in the throat that it was our own guys who'd been involved.

The eight men involved in the operation walked free because the arrest was made 900 miles outside of territorial waters. The defence said it was like 'arresting a shoplifter before he got to the checkout' even though the ship showed all the evidence of it heading for the UK; for instance, her GPS Way Point route led directly to England. The customs team was criticised for their lack of judgement: boarding the ship was classed as illegal because she was in international waters.

The eight men appealed and were then released – absolutely no justice. And the media portrayed these men as SBS operators. They were not: they were 'attached ranks', are still classed as 'turncoats' and will never be able to rebuild their lives within the Royal Marines or SBS.

Glossary

AC	Air Coalition
AO	area of operation
ASASR	Australian Special Air Service Regiment
ASW	anti-submarine warfare
BAe	British Aerospace
BATT	British Advisory Training Team
Bergen	backpack
Buddy line	a line attached to a swimming partner and used as a measuring distance
CO	Commanding Officer
CQR	Canadian Quick Release
CQB	Close Quarter Battle
CT	Counter Terrorism
CTCRM	Command Training Centre Royal Marines
the Det	14 Intelligence Company
DLO	Drugs Liaison Officer
DOP	drop-off point
DSF	Director Special Forces
EMCON	Emission Control
ER	emergency response
E & E	escape and evasion
E & RE	exit and re-entry
EXFIL	exfiltration
FFD	First Field Dressing

FMB	Forward Mounting Base
FOB	Forward Operating Base
14 Int.	14 Intelligence Company
FRV	final rendezvous
Gemini	inflatable assault boat
GPMG	General Purpose Machine Gun
H hour	time for attack
Head Shed	headquarters
HK	Heckler & Koch
HKMP5SD	Heckler & Koch MP5 SD (Silenced Version)
HKMP5K	Heckler & Koch MP5 A1 (Short Barrel – No Extendable Butt Version)
HKMP5A3	Heckler & Koch MP5 A3 (Long Barrel – Extendable Butt Version)
Karabiner	a metal clip with a spring
Klepper	canoe
LAR 5	attack oxygen breathing set
LAW	Light Anti-Tank Weapon
LCU	Landing Craft Utility
LIFA	light, warm, breathing garment
LMF	low moral fibre
LRIC	Long Range Insertion Craft
LS	Landing Sight (Single Aircraft)
LUP	lying-up position
Lurking Area	Waiting Area on the Casing of an 'O' Boat Submarine
LZ	Landing Zone (Multiple Aircraft)
MAT	Maritime Anti-Terrorism
MB	Mounting Base
MCT	Maritime Counter Terrorism
ML	Royal Marines Mountain Leader
MoD	Ministry of Defence
MRE	meals ready to eat
MSD	Maximum Safety Depth
NATO	North Atlantic Treaty Organisation
NCIS	National Crime Intelligence Service

NCO	non-commissioned officer
NGs	night goggles
NVG	night viewing goggles
O boat	diesel submarine
on stag	sentry duty
OP	observation post
Op	operation
Opsec	operational security
PADI	Professional Association of Diving Instructors
pepper-pot	one man moves forward while another man covers him
pulk	stores ski sledge
RABA	re-chargeable air breathing apparatus
Red Area	Maximum-security area
Re-sup	resupply
RIB	Rigid Inflatable Boat
Rigid Raiders	Rigid Fast Boats
RN	Royal Navy
RSM	Regimental Sergeant Major
RTI	resistance to interrogation
RTU	return to unit
RV	rendezvous
SARBE	Search and Rescue Beacon Emergency
SAS	Special Air Service
SBS	Special Boat Service
SCBA	Swimmer Canoeist Breathing Apparatus (Oxygen)
SCLJ	Swimmer Canoeist Life Jacket
SF	Special Forces
SLR	Self-Loading Rifle
SNCO	senior non-commissioned officer
SOP	Standard Operating Procedure
SSM	Squadron Sergeant Major
Taps and Bells	A series of knocks to indicate intention
TLZ	Total Exclusion Zone
VCP	vehicle check point

Woolly Bears	thick thermal suits
Yomp	a long walk with heavy weights
Zodiac	Inflatable Boat